Magnetic ceramics describes the structure, preparation techniques, magnetic properties and applications of iron-based oxides, also known as ferrites. The main purpose of the book is to provide an up-to-date overview of the relevant aspects of ferrites. These cover a wide range of magnetic properties and applications such as high-frequency transformer cores, permanent magnets, microwave telecommunication devices, magnetic recording tapes and heads. The subject is approached by emphasising its interdisciplinary nature. Synthesis and fabrication methods and crystal structures are covered alongside the theoretical basis of magnetic properties. This book fills a gap in the ceramics literature. It has been written to appeal to the international community of materials scientists, including researchers in industry and graduate students.

Magnetic ceramics

Chemistry of Solid State Materials

Series editors:
Bruce Dunn, Department of Materials Science and Engineering, UCLA
John W. Goodby, School of Chemistry, University of Hull
A. R. West, Department of Chemistry, University of Aberdeen

Magnetic ceramics

Raul Valenzuela

Instituto de Investigaciones en Materiales,
National University of Mexico

CAMBRIDGE
UNIVERSITY PRESS

CAMBRIDGE UNIVERSITY PRESS
Cambridge, New York, Melbourne, Madrid, Cape Town, Singapore, São Paulo

Cambridge University Press
The Edinburgh Building, Cambridge CB2 2RU, UK

Published in the United States of America by Cambridge University Press, New York

www.cambridge.org
Information on this title: www.cambridge.org/9780521364850

First published 1994
This digitally printed first paperback version 2005

A catalogue record for this publication is available from the British Library

Library of Congress Cataloguing in Publication data
Valenzuela, Raul.
Magnetic ceramics / Raul Valenzuela.
 p.   cm. — (Chemistry of solid state materials)
Includes bibliographical references.
ISBN 0-521-36485-X
1. Ceramic materials—Magnetic properties.   2. Ferrites—Magnetic
properties.   I. Title.   II. Series.
TK7871.15.C4V35   1994
620.1′404297—dc20   93-43325   CIP

ISBN-13 978-0-521-36485-0 hardback
ISBN-10 0-521-36485-X hardback

ISBN-13 978-0-521-01843-2 paperback
ISBN-10 0-521-01843-9 paperback

Contents

Contents

Contents

Preface

Excellent reviews of various aspects of ferrites (Smit & Wijn, 1961; Smit, 1971; Wohlfarth, 1980, 1982; Goldman, 1990) and magnetism (Cullity, 1972; Jiles, 1991) have been published; however, the need for a monograph covering recent developments on ferrites, such as new preparation technologies, the now-established correlations between polycrystalline structure and magnetic properties, and the most recent applications in magnetic recording, at an introductory level, was felt. The aim of this book is to satisfy that need.

The book corresponds to a final-year undergraduate or graduate level in solid-state chemistry, solid-state physics, or materials science, but is expected to be useful also for the researcher specialising in other areas who is interested in a basic overview of magnetic ceramics. It is hoped, additionally, that it will be helpful for the development engineer concerned with the science of ferrites which underpins their applications.

The book was conceived to cover all the important aspects of ferrites. The different crystalline structures are described in Chapter 2, with brief discussions on cation site occupancy and the basic features of the polycrystalline state. Chapter 3 is devoted to preparation methods for powders, ceramics, thin films and single crystals, including classical solid-state sintering as well as new techniques involving the manipulation of small particles. Magnetic properties are discussed in Chapter 4; the approach is developed from the microscopic scale (magnetic moments localised on atoms) to the macroscopic level (domains, granular and shape effects). Electrical and magnetooptic properties are also briefly discussed. Chapter 5 describes the basis of the well-established applications such as permanent magnets and high-frequency transformers, more complex devices based on ferrites, such as magnetic recording heads and media, as well as some of the most promising developments, such as perpendicular recording media and magnetooptic recording materials. Finally, a brief overview of other types of magnetic materials, mostly metallic alloys, is given in Chapter 6, including 'classic' materials such as soft and hard magnets, magnetic recording media and heads, miscellaneous materials and applications, and superconductors. In each chapter, an extensive list of references is provided for readers interested in having access to the full details of the original papers.

SI-Sommerfeld units are used throughout this book, since it is essentially addressed to newcomers to magnetism. It is expected that the SI unit system will eventually be fully adopted by the magnetism research community. The definitions of the basic parameters and the conversion factors are given in Tables 4.6–4.8.

Most of this book was written during a sabbatical leave at the Department of Chemistry, University of Aberdeen, with a Research Fellowship from the European Commission; I thank the Mexican authorities at SRE and at the University of Mexico (DGAPA-UNAM). I especially acknowledge the invitation and encouragement from Professor Anthony West to write this book, as well as his invaluable help with the intricacies of the English language! I am indebted to the staff of the IIM-UNAM (Mexico) and Queen Mother (Aberdeen) Libraries for providing many scientific publications. I thank my wife Patricia, not only for her understanding and encouragement to achieve this task, but also for her technical help (she drew most of the figures!). Finally, I hope that my son Alejandro will be interested in this book in the near future.

University of Mexico   Raúl Valenzuela
July 1993

References

Cullity, B. D. (1972). *Introduction to Magnetic Materials*, Addison-Wesley, Massachusetts.

Goldman, A. (1990). *Modern Ferrite Technology*, Van Nostrand Reinhold, New York.

Jiles, D. (1991). *Introduction to Magnetism and Magnetic Materials*, Chapman and Hall, London.

Smit, J. (1971). *Magnetic Properties of Materials*, McGraw-Hill, New York.

Smit, J. & Wijn, H. P. J. (1961). *Les Ferrites*, Bibliotheque Technique Philips, Dunod, Paris.

Wohlfarth, E. P. (Ed.) (1980). *Ferromagnetic Materials*, Vol. 2. North-Holland, Amsterdam.

Wohlfarth, E. P. (Ed.) (1982). *Ferromagnetic Materials*, Vol. 3. North-Holland, Amsterdam.

Symbols

A	exchange constant
A, A'	constant
A	cross-sectional area
A	angular momentum
a	nucleation-rate factor
a	unit cell parameter
$\mathbf{B}(J, \alpha')$	Brillouin function
B	magnetic induction
B_d	demagnetised induction
B_r	remanent induction
$(BH)_{max}$	maximum energy product
b	growing phase dimensionality factor
C	capacitance
C	domain wall mobility
C_o, C_c	vacancy concentrations
C_v	heat capacity at constant volume
c	cell parameter
D	divalent cation
D	average grain size
D_0, D_c	diffusion coefficients
Dq	energy difference (crystal field)
d_{xy}	atomic orbitals
d	density
E	activation energy
E_c	kinetic energy
E_f	Fermi energy
E_K	anisotropy energy
E_{ex}	exchange energy
E_m	magnetostatic energy
E_n	electron energy
E_p	magnetic potential energy
e	electron charge
e_g	electron orbital doublet
$\mathbf{F}$	Lorentz force

Symbols

f	frequency
g	Landé factor
H	intensity of magnetic field
H_c	coercivity
H_c	critical field (superconductors)
H_{c_1}, H_{c_2}	critical fields (superconductors)
H_{cr}	critical field
H_{eff}	effective anisotropy field
$_B H_c$	coercive field measured from B–H curves
$_i H_c$	intrinsic coercive field
H_K	anisotropy field
H_n	nucleation field
H_p	pinning field
H_p	propagating field
H_T	total field
H_W	molecular or Weiss field
ΔH	resonance linewidth
h	ac magnetic field
h	Planck's constant
h_{rf}	radio frequency magnetic field
$I(K)$	interference function
J	total angular quantum number
J_{ex}	exchange integral
K	Arrhenius-type constant
K	geometric constant
K	Kundt's constant
K	total anisotropy constant
K	scattered density
K_s	shape-anisotropy constant
K_1, K_2	magnetocrystalline anisotropy constants
k	constant
k	Boltzmann constant
L	inductance
L	total angular quantum number
$L(\alpha)$	Langevin function
l	angular momentum quantum number
M	magnetisation
M	metaloid
Me	divalent cation
M_r	remanent magnetisation
M_s	saturation magnetisation
$\mathscr{M}$	magnetic dipolar momentum
m	reduced magnetisation
m	mass
m	effective domain wall mass
m_l	magnet quantum number

Symbols

N	number of atoms per volume unit
N_d	demagnetisation factor
N_0	Avogadro's number
n	Avrami exponent
n	integer
n	number of Bohr magnetons per formula unit
n	total number of electrons 3d + 4s
P_e	eddy-current loss
P_h	hysteresis energy loss
$P(r)$	atomic distribution function
P_T	total energy loss
P_0	average atomic density
ΔP	change in properties
ΔP	pressure difference
p	growth mechanism factor
p_s	spin angular momentum
p_0	angular momentum
Q	electric charge
q	electric charge
R	gas constant
R	rare-earth cation
R	resistance
R_0	oxide ion radius
$RDF(r)$	radial distribution function
R	wave function in spherical coordinates
r	distance between atoms
r	heating rate
r	particle radius
r	pore radius
r_0	interatomic equilibrium distance
S	total spin
s	spin quantum number
s_p	particle separation
T	temperature
T	trivalent cation
TM	transition metal
T_C	Curie temperature
T_f	fictive temperature
T_g	glass transition temperature
T_m	melting temperature
T_N	Néel temperature
T_x	crystallisation temperature
T_{comp}	compensation temperature
t	thickness
t	reduced temperature
t	time

Symbols

t_{2g}	electron orbital triplet
u	deformation parameter
V	unit cell volume
V	electrostatic potential
$\mathbf{v}$	velocity
v	volume
x	composition
x	crystallised fraction
x	domain wall displacement
W	formula weight
W_h	area of hysteresis loop
Z	atomic number
Z^*	complex impedance
α	Fe bcc crystal phase
α	Langevin factor
α	optical absorption coefficient
α	restoring force coefficient
$\alpha_1, \alpha_2, \alpha_3$	angle cosines
β	viscous damping factor
γ	gyromagnetic ratio
γ	surface energy
γ	Fe fcc crystal phase
γ_w	domain wall energy
Δ	energy difference (crystal field)
δ	degree of inversion
δ_w	domain wall thickness
ε	permittivity
η	geometrical factor
Θ	extrapolated temperature (antiferromagnetics)
Θ	wave function in spherical coordinates
θ	angle
θ_F	Faraday rotation angle
λ	molecular field coefficient
λ	wavelength
λ_s	saturation magnetostriction constant
μ_i	relative initial permeability
μ_{max}	relative maximum permeability
μ_0	permeability of free space
ρ	electrical resistivity
σ_g	greenbody strength
τ	relaxation time
Φ	wave function in spherical coordinates
Φ_0	magnetic quantum flux
ϕ	magnetic flux
ϕ_v	solid volume fraction

xviii

Symbols

χ	magnetic susceptibility
χ^*	complex susceptibility
χ', χ''	real and imaginary parts of the susceptibility
χ_0	low-frequency susceptibility
$\chi_\perp$	perpendicular magnetic susceptibility
$\chi_\parallel$	parallel magnetic susceptibility
Ψ	wave function
ψ	time-independent wave function
ω	angular frequency
ω_L	Larmor frequency
ω_s	resonance frequency
ω_x	angular relaxation frequency

1 Introduction

We are surrounded by magnetic materials playing a crucial role in many devices of every-day life: ac and dc motors which perform many operations (a top-of-the-range car has more than 20 dc motors); power distribution systems, based on power transformers, which deliver energy for home and industrial use; video and audio applications (tapes, writing/reading heads) which provide information and entertainment on a massive scale; telephone and telecommunication systems (microwave devices) which link continents at nearly the speed of light; data storage systems (discs, disc drives) which pervade virtually every human activity.

The history of magnetic materials is as old as that of man; the strange properties of magnetite (a magnetic ceramic!) have been linked to the military successes of an ancient Chinese Emperor (Huang-Ti, 2600 BC). The word magnet is derived from a Greek word used to indicate magnetite deposits in the district of Magnesia. The first study of magnetism is the book *De Magnete*, by W. Gilbert, published in 1600. The next significant development in magnetism occurred in 1825, when H. C. Oersted reported the crucial fact that magnetic fields can be produced by means of electrical currents. This discovery opened the way to the first applications of magnetism. Discoveries, models and theories have developed since at an increasingly accelerated pace. An amazing fact is that all the impressive variety of magnetic materials and their properties originate mainly from the three elements that are ferromagnetic at room temperature: iron, cobalt and nickel.

Magnetic ceramics, or ferrites, are a very well-established group of magnetic materials. Research activity on ferrites, especially intense in the last 50 years, has led to the establishment of many theories and models additional to or complementing those obtained from research on metallic materials. Magnetic ceramics participate in virtually every application area; in some cases, there are no other practical alternative materials.

1

Introduction

Ferrites have become a reference material; developments associated with other new magnetic materials, such as the extra-hard rare-earth inter-metallics (Buschow, 1990), or the extra-soft amorphous ribbons (Boll & Hilzinger, 1983) are often assessed by comparison with ferrites.

Magnetic ceramics are well established, but improvements and innova-tions continue to take place; many new and exciting applications, theories and preparation technologies are currently under development. For instance, at the last scientific meeting devoted entirely to ferrites (held every four years), the 6th International Conference on Ferrites, Japan, October 1992, more than 550 research papers were presented by some 1159 authors.

Ferrites are complex because they combine two complex areas: ceramic microstructures and magnetic phenomena. Ceramic microstructures, formed as a result of physico-chemical processes such as solid-state sintering, are affected by a large number of interacting variables; the essentially quantum-mechanical nature of their magnetic properties makes them difficult to comprehend, since they are entirely different to macroscopic, every-day experience. The approach to ferrites; their synthesis/fabrication; the relationship between crystal structure, texture and physical properties; the modelling of magnetic interactions, is of necessity interdisciplinary.

In this book, an attempt is made to provide an overview of the science of magnetic ceramics. Chemical aspects are covered in terms of synthetic methods and crystallography. Physics is introduced to provide a theoretical basis to magnetism, which is necessary to interpret the property measure-ments. Materials science links together physics and chemistry and, in addition, provides the framework for a scientific understanding of fabrication and testing, leading to applications.

References

Boll, R. & Hilzinger, H. R. (1983). Comparison of amorphous materials, ferrites and permalloys. *IEEE Transactions on Magnetics*, **Mag-19**, 1946–51.
Buschow, K. H. J. (1990). New Developments in hard magnetic materials. *Reports of Progress in Physics*, **54**, 1123–213.

2 The crystal structures of magnetic ceramics

2.1 *Spinels*

2.1.1 *The spinel structure*

The *spinel* ferrites are a large group of oxides which possess the structure of the natural spinel $MgAl_2O_4$. More than 140 oxides and 80 sulphides have been systematically studied (Hill, Craig & Gibbs, 1979). Many of the commercially important spinels are synthetic, but one of the most important and probably the oldest magnetic material with practical applications, magnetite Fe_3O_4, is a natural oxide.

Their great abundance points to a very stable crystal structure. Spinels are predominantly ionic. The particular sites occupied by cations are, however, influenced by several other factors, including covalent bonding effects (e.g., Zn in tetrahedral sites) and crystal field stabilisation energies of transition-metal cations.

Many different cation combinations may form a spinel structure; it is almost enough to combine any three cations with a total charge of eight to balance the charge of the anions. The limits of the cation radii are approximately 0.4–0.9 Å (based on the oxide ion radius, R_o, of 1.4 Å). The following combinations are known:

2, 3	as in	$NiFe_2O_4$
2, 4	as in	Co_2GeO_4
1, 3, 4	as in	$LiFeTiO_4$
1, 3	as in	$Li_{0.5}Fe_{2.5}O_4$
1, 2, 5	as in	$LiNiVO_4$
1, 6	as in	Na_2WO_4

The most important spinels from the magnetic point of view are the oxides 2, 3. Some of the sulphides also have interesting magnetic properties

3

(Lotgering, 1956) but their magnetic ordering, which is the important factor for applications, is observed only at low temperatures and therefore they are unsuitable for applications at room temperature. A particular case is maghemite, γ-Fe_2O_3, which possesses a defect spinel structure. The Fe cations are in the trivalent state, and a fraction of the cation sites is vacant (Verwey, 1935).

The spinel structure was first determined by Bragg (1915) and Nishikawa (1915). The ideal structure is formed by a cubic close-packed (fcc) array of O atoms, in which one-eighth of the tetrahedral and one-half of the octahedral interstitial sites are occupied by cations. The tetrahedrally coordinated sites and the octahedrally coordinated sites are referred to as the A and B sites, respectively.

The unit cell contains eight formula units AB_2O_4, with eight A sites, 16 B sites and 32 oxygens. It can be described (Smit & Wijn, 1961) by taking an A site as the origin of the unit cell. It is convenient to divide the unit cell into eight cubes of edge $a/2$ to show the arrangement of the A and B sites, Fig. 2.1. The space group is Fd3m. The O atoms have a four-fold coordination, formed by three B cations and one A cation, Fig. 2.2.

The O atoms in the spinel structure are not generally located at the exact positions of the fcc sublattice. Their detailed positions are determined by a parameter, u, which reflects adjustments of the structure to accommodate differences in the radius ratio of the cations in the tetrahedral and octahedral sites. The u parameter is defined in Fig. 2.3 and has a value of 0.375 for an ideal close-packed arrangement of O atoms, taking as unit cell that of Fig. 2.1. An alternative definition of this parameter can be given (Hill, Craig & Gibbs, 1979) by using the centre of symmetry, located at (0.125, 0.125, 0.125), as the origin of the unit cell. In this case, the ideal u value is 0.25. In the following, the tetrahedral-based definition with an A site at the origin will be used.

The ideal situation is almost never realised, and the u value for the vast majority of the known spinels ranges between 0.375 and 0.385. u increases because the anions in tetrahedral sites are forced to move in the [111] direction to give space to the A cations, which are almost always larger than the ideal space allowed by the close-packed oxygen, but without changing the overall $\overline{4}$3m symmetry. Octahedra become smaller and assume 3m symmetry. In Table 2.1, interatomic distances are given as a function of the unit cell parameter a and the u parameter.

The average radii of the cations affect primarily the cell parameter a,

Table 2.1. *Interatomic distances and site radii in spinels* AB_2O_4, *as a function of unit cell edge a and deformation parameter u.*

Tetra–tetra separation A–A	$a\sqrt{3}/4$
Tetra–octa separation A–B	$a\sqrt{11}/8$
Octa–octa separation B–B	$a\sqrt{2}/4$
Tetra–O separation A–O	$a\sqrt{3}(u - 0.25)$
Octa–O separation B–O	$a[3u^2 - 2.75u + 43/64]^{1/2} \sim a(\frac{5}{8} - u)$
O–O tetrahedral edge O–O	$a\sqrt{2}(2u - 0.5)$
O–O shared octa edge O–O	$a\sqrt{2}(1 - 2u)$
O–O unshared octa edge O–O	$a[4u^2 - 3u + \frac{11}{16}]^{1/2}$
Tetrahedral radius	$a\sqrt{3}(u - 0.25) - R_o$
Octahedral radius	$a[3u^2 - 2.75u + \frac{43}{64}]^{1/2} - R_o \sim a(\frac{5}{8} - u) - R_o$

u is defined with the unit cell origin at an A site; R_o is the oxide ion radius. (Adapted from Hill, Craig & Gibbs, 1979, and Jagodzinski, 1970.)

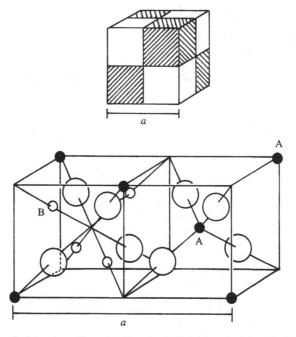

Fig. 2.1. The spinel structure. The unit cell can be divided into octants; tetrahedral cations A, octahedral cations B, and O atoms (large circles) are shown in two octants. (Adapted from Smit & Wijn, 1961.)

while the ratio between the tetrahedral and octahedral cation radii determines mainly the u value.

If the lattice parameter is taken as a weighted average of the projections of the octahedral and tetrahedral bond lengths on the unit cell, the lattice parameter can be approximated by the expression:

$$a_{\text{calc}} = \frac{8(\text{tet bond})}{3\sqrt{3}} + \frac{8(\text{oct bond})}{3} \tag{2.1}$$

This expression accounts for 96.7% of the variations in the lattice parameter of 149 spinel oxides (Hill, Craig & Gibbs, 1979).

2.1.2 Normal and inverse spinels: cation distribution

In the spinel $MgAl_2O_4$, Al and Mg cations occupy the octahedral and tetrahedral sites, respectively. This cation distribution is usually indicated

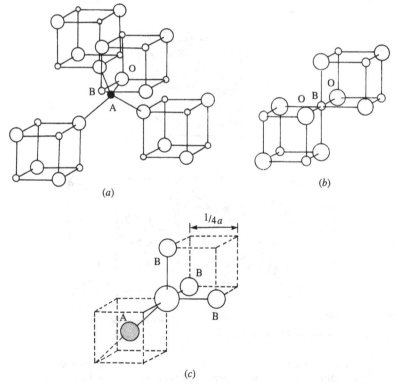

Fig. 2.2. Nearest neighbours of (*a*) a tetrahedral site, (*b*) an octahedral site and (*c*) an anion site. (Adapted from Smit & Wijn, 1961.)

as:

$$(Mg)[Al_2]O_4 \qquad\qquad (2.2)$$

where the square brackets indicate the octahedral site occupancy, and the cations in parentheses are located in the tetrahedral sites. This is the so-called *normal* distribution. If D denotes a divalent cation, and T a trivalent one, another extreme cation distribution is:

$$(T)[DT]O_4 \qquad\qquad (2.3)$$

which is called an *inverse* spinel. In many cases, intermediate cation distribution has been observed, i.e.:

$$(D_{1-\delta}T_\delta)[D_\delta T_{2-\delta}]O_4 \qquad\qquad (2.4)$$

where δ is the *degree of inversion,* with a value of zero for the normal and one for the inverse distributions, respectively. In many cases, the degree of inversion depends on the preparation technique, especially on the cooling rate after sintering.

The physical properties of spinels depend not only on the kinds of cation in the lattice, but also on their distribution over the available crystal

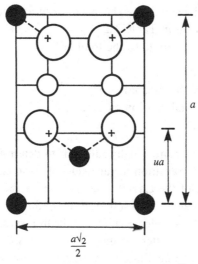

Fig. 2.3. Definition of the deformation parameter u. Half of a $(\bar{1}10)$ plane is shown. O atoms (large circles) are displaced along the $\langle 111 \rangle$ directions due to the presence of tetrahedral cations. In the ideal, undeformed case, $u = 0.375$. (Adapted from Smit & Wijn, 1961.)

Table 2.2. *Cation distribution, lattice parameter and u value for several spinels.*

	Distribution	a (Å)	u
Normal	$(Cd)[Fe_2]$	8.7050	0.3935
	$(Zn)[Fe_2]$	8.4432	0.3865
Inverse	$(Fe)[CoFe]$	8.3500	0.381
	$(Fe)[CuFe]^a$	8.3690	0.380
	$(Fe^{3+})[Fe^{2+}Fe^{3+}]$	8.3940	0.3798
	$(Fe)[Li_{0.5}Fe_{1.5}]$	8.3300	0.3820
	$(Fe)[NiFe]$	8.3390	0.3823
Mixed	$(Mg_{1-\delta}Fe_\delta)[Mg_\delta Fe_{2-\delta}]$	8.3600	0.3820 ($\delta = 0.1$)
	$(Mn_{1-\delta}Fe_\delta)[Mn_\delta Fe_{2-\delta}]$	8.5110	0.3865 ($\delta = 0.85$)
	$(Mo_{1-\delta}Fe_\delta)[Mo_\delta Fe_{2-\delta}]$	8.5010	0.3751 ($\delta = 0.5$)

^{a} Below 760 °C, this spinel has a tetragonal deformation, with $a = 8.70$ Å and $c = 8.22$ Å (Prince & Treuting, 1956).

sites. It is thus of major importance to understand what factors influence the site occupancy. In fact, understanding and predicting the cation distribution in spinels have been among the more interesting and persistent problems in crystal chemistry. The experimental cation distribution of selected ferrites is given in Table 2.2.

The factors that contribute to the total lattice energy in spinels are:

(1) elastic energy;
(2) electrostatic (Madelung) energy;
(3) crystal field stabilisation energy;
(4) polarisation effects.

The first two energies are usually sufficient to determine the total lattice energy in ionic, non-transition-metal oxides. Elastic energy refers to the degree of distortion of the crystal structure due to the difference in ionic radii assuming that ions adopt a spherical shape. Smaller cations, with ionic radii of 0.225–0.4 Å, should occupy tetrahedral sites, while cations of radii 0.4–0.73 Å should enter octahedral ones. This distribution leads to a minimum in lattice strain. Since trivalent cations are usually smaller than divalent ones, a tendency toward the inverse arrangement would be expected in 2, 3 spinels.

Detailed calculations of the Madelung energy for spinels (Verwey, de Boer & van Santen, 1948) show that this energy is dependent on the u parameter. For $u > 0.379$, the normal distribution is more stable, while for lower u values, the inverse arrangement possesses a higher Madelung constant. The presence of two kinds of cation on octahedral sites in inverse spinels leads to an additional contribution to the Madelung energy if long-range order of these B-site cations is established, Fig. 2.4. The critical u value then becomes 0.381 (de Boer, van Santen & Verwey, 1950); electrostatic energy is higher for the normal spinel if $u > 0.381$, and the inverse, ordered spinel is more stable for $u < 0.381$.

Experimental results are not in agreement with these calculations (Table 2.2), which is not surprising. These simple estimates of the elastic and electrostatic contributions are based on the assumption of spherically symmetric ions with only coulombic interactions, which is far from being the case for transition-metal cations in ferrimagnetic spinels.

The application of *crystal field* theory to the understanding of cation site 'preference' was first suggested by Romeijn (1953). As is well known, the charge density of d orbitals, Fig. 2.5, interacts with the charge distribution of the environment in which the transition ion is placed.

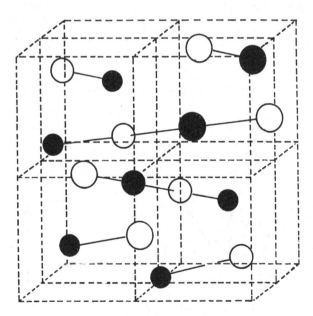

Fig. 2.4. 1:1 long-range order in inverse spinels, resulting from the alternating occupancy of octahedral sites by two types of cation. Only octahedral cations are shown (Aburto *et al.*, 1982).

The five d orbitals (commonly designated as d_{xy}, d_{yz}, d_{zx}, d_{z^2} and $d_{x^2-y^2}$) no longer have the same energy, but are split according to the symmetry of the electrostatic field produced by the anions of the particular lattice site. The physical basis for this splitting is simply the electrostatic repulsion between the d electrons and the electrons of the orbitals of the surrounding anions. The group-theoretical aspects of crystal field theory are reviewed by Cracknell (1975).

In an octahedral field, the energy level splitting leads to two groups of orbitals, a lower triplet formed by the d_{xy}, d_{yz} and d_{zx} orbitals, and a higher doublet with the d_{z^2} and $d_{x^2-y^2}$ orbitals, Fig. 2.6. The energy of the doublet is increased as these orbitals point directly to the anions, while the triplet energy decreases, because the orbitals point to regions of low electron density. The energy difference between the triplet and the doublet is given as $10Dq$, or Δ.

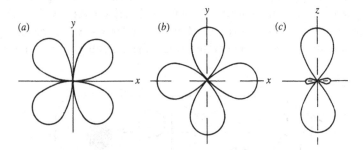

Fig. 2.5. Spatial geometry of d orbitals: (*a*) d_{xy}, (*b*) $d_{x^2-y^2}$ and (*c*) d_{z^2}.

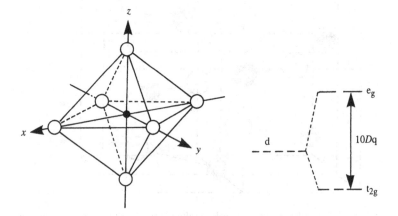

Fig. 2.6. Transition-metal cation in an octahedral field and its schematic energy level diagram.

Table 2.3. *Crystal field stabilisation energies for transition-metal cations on octahedral and tetrahedral spinel sites.*

Number of d electrons	Theoretical cfs in terms of Dq		Cations	Estimated[a] octahedral site preference energy, eV
	Octahedral	Tetrahedral		
1	4	6	Ti^{3+}	0.33
2	8	12	V^{3+}	0.53
3	12	8	V^{2+}	1.37
			Cr^{3+}	2.02
4	6[b]	4	Mn^{3+}	1.10
			Cr^{2+}	0.74
5	0	0	Fe^{3+}	0
			Mn^{2+}	0
6	4	6	Fe^{2+}	0.17
			Co^{3+}	0.82
7	8	12	Co^{2+}	0.09
8	12	8	Ni^{2+}	0.99
9	6[b]	4	Cu^{2+}	0.68
10	0	0	Zn^{2+}	0

[a] From McClure (1957).
[b] Jahn–Teller additional stabilisation.

Hund's rules state (Condon & Shortley, 1935) that 'the electron states with the greatest $(2S + 1)$ are most stable, and of those, the most stable is that with the greatest L.' This leads to a tendency to the high spin state (HS), i.e., with the highest number of unpaired electron spins. For d^1, d^2 and d^3 cations in octahedral sites, the lower energy triplet is occupied by unpaired electrons; the d^3 state (V^{2+}, Cr^{3+}, Mn^{4+}) has the highest stabilisation energy, see Table 2.3. Ni^{2+} with the d^8 configuration (the triplet occupied by six paired electrons and the doublet by two unpaired electrons) will also be particularly stable in octahedral coordination.

In the case of tetrahedral sites, the splitting is reversed; the doublet has a lower energy than the triplet, Fig. 2.7. The energy difference in tetrahedrally coordinated cations is a fraction ($\frac{4}{9}$) of that for the octahedral coordination. Half-filled d orbitals (Mn^{2+} and Fe^{3+}) in the high spin state have a d^5 spherical configuration with no particular preference for either coordination.

Several quantitative studies (Dunitz & Orgel, 1957; McClure, 1957) have allowed a direct evaluation of the importance of this contribution to the cation distributions in spinels. By combining spectroscopic and magnetic data on a variety of compounds, an estimate of the octahedral site preference energy was obtained. This is the difference in crystal field stabilisation energy between octahedral and tetrahedral sites, and is given in Table 2.3.

If the crystal field splitting energy is very large, it becomes more favourable to fill the lower triplet orbitals, thereby violating Hund's rules and decreasing the total spin. This is the *low-spin* configuration, Fig. 2.8. For example, the Rh^{3+} ($4d^6$) ion in octahedral coordination may adopt the $(t_{2g})^6$ configuration (i.e., with the six electrons paired in the lower triplet), with zero total magnetic moment. The *high-spin* state is $(t_{2g})^4(e_g)^2$, which can be schematically indicated as

$$\boxed{\uparrow\downarrow}\ \boxed{\uparrow}\ \boxed{\uparrow}$$

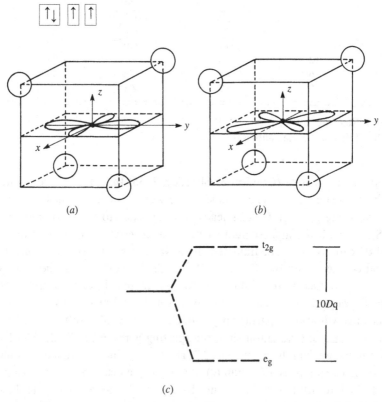

Fig. 2.7. Transition-metal cation in a tetrahedral coordination; (a) $d_{x^2-y^2}$ orbital, (b) d_{xy} orbital and (c) corresponding schematic energy level diagram.

in the lower triplet, and

in the doublet. The total spin moment is therefore $S = 2$.

It turns out that crystal field theory accounts adequately for practically all the experimental results in spinels. The fact that all the known chromites have a normal distribution is consistent with the high octahedral field stabilisation energy value calculated for Cr^{3+} ($S = 3/2$). In ferrites, the arrangement is very dependent on the divalent cation, since Fe^{3+} has no crystal field stabilisation energy. When the divalent cation also shows no clear preference, ferrites with δ values between zero and one (*mixed ferrites*) are obtained.

It may at first sight seem improbable that crystal field stabilisation theory should be able to account for cation distributions when its magnitude (about 2 eV for Cr^{3+}, the highest value in transition-metal cations) is compared with that of other contributions, especially the Madelung energy. For the latter, a variation in the u parameter as small

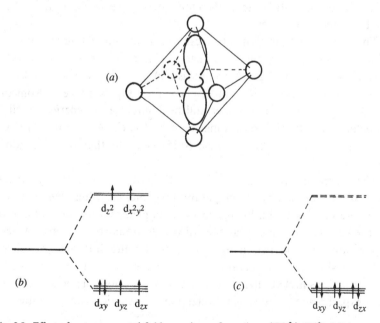

Fig. 2.8. Effect of a strong crystal field on spin configuration of Rh^{3+} ($4d^6$): (a) d_{z^2} orbital in an octahedral site; (b) high-spin configuration (weak crystal field); and (c) low-spin configuration (strong crystal field).

as 0.005 involves an electrostatic energy variation of 4 eV (Verwey & Heilmann, 1947). The u variations between a normal and an inverse distribution are certainly greater than 0.005, but it appears as if the lattice parameters accommodate these differences, leading to a practically unchanged total electrostatic energy. d orbital splitting therefore becomes the dominant influence. As noted in Table 2.3, d^4 and d^9 cations may lead to an additional stabilisation effect, through the Jahn–Teller distortion. This is discussed in Section 2.1.4.

The last factor to be discussed is polarisation effects. Polarisation may simply be considered as the degree of distortion of the electronic charge density around an ion, and can arise from many causes. The two extreme cases, i.e., negligible distortion and effective removal of an electron from one ion toward its neighbour, give rise to a purely covalent bond and a purely ionic bond, respectively. With regard to transition-metal ions in spinels, it is expected that only spherically symmetric ions (d^5 and d^{10}) can show a tendency for covalency. In this case, tetrahedral sites are preferred. Cations which show covalent affinity for tetrahedral environments are Fe^{3+}, Ga^{3+}, In^{3+} and, more strongly, Zn^{2+} and Cd^{2+}. Spinels with the former cations tend therefore to be inverse, while those with the latter tend to be normal.

The actual configuration is the result of many interacting factors; very often, the difference in the total energy of normal versus inverse distributions is very small. A good example is $NiAl_2O_4$. The expected Madelung energy for the normal distribution, $(Ni)[Al_2]O_4$, is 1.08 eV (Romeijn, 1953); on the other hand, the crystal field stabilisation energy for Ni^{2+} on octahedral sites (inverse arrangement) is 0.91 eV. As a result, $NiAl_2O_4$ is a mixed spinel, with a δ value (0.75) close to that of the random distribution (0.67).

The degree of inversion, δ, is not exactly zero (or one) in many ferrites, and can be modified by the preparation technique. At high temperatures (near the melting point, for instance), a simple statistical distribution of all the cations on the crystal sites ($\delta = 2/3$) would be expected. A very rapid cooling can quench-in high-temperature states. If the cation redistribution is slow, high-temperature states remain at room temperature. In many cases, a simple Boltzmann distribution has shown a good agreement with experimental results. The Boltzmann distribution can be written:

$$\frac{(2-\delta)(1-\delta)}{\delta^2} = e^{\frac{-E}{kT}} \tag{2.5}$$

where E is the activation energy for the site exchange between a divalent cation and a trivalent one, k is the Boltzmann constant and T the temperature. The *thermodynamics* of cation redistribution is well understood; by varying the cooling rate between 0.01 °C/s and 1000 °C/s, a continuous variation in cation configuration is obtained in $CoFe_2O_4$ (De Guire, O'Handley & Kalonji, 1989). A typical value of E is 0.8 ± 0.05 eV for an inverse ferrite $NiFe_2O_4$ (Robertson & Pointon, 1966), and 0.14 eV for a mixed ferrite, $MgFe_2O_4$ (Bacon & Roberts, 1953). The cation distributions of the following ferrites have been investigated by neutron diffraction: $NiFe_2O_4$ and $ZnFe_2O_4$ (Hastings & Corliss, 1953), $MgFe_2O_4$ (Corliss, Hastings & Brockman, 1953), $MnFe_2O_4$ (Hastings & Corliss, 1956), and Fe_3O_4 (Shull, Wollan & Koehler, 1951).

These mechanisms of cation redistribution (the *kinetics*), however, are more complex and can be significantly affected by the presence of Fe^{2+} (Sujata & Mason, 1992). When a substantial concentration of divalent Fe was present in an inverse spinel, the redistribution kinetics were independent of the cation vacancy concentration, grain size and O stoichiometry. Ferrites containing Fe in the trivalent state only, such as $MgFe_2O_4$, have shown instead a clear dependence on grain size and other factors associated with nucleation and growth mechanisms (Kimura, Ichikawa & Yamaguchi, 1977).

In the case of inverse ferrites (or in any case where more than one type of cation occupies the same set of crystallographic sites), *ordering* phenomena can be expected (Gorter, 1954). In inverse ferrites, octahedral site occupancy by two kinds of cations (divalent and trivalent) can lead to long-range order, where successive (001) layers of octahedral sites are occupied alternately by D and T cations. In this case, there are two ionic sublattices (Fuentes, Aburto & Valenzuela, 1987) on octahedral sites, Fig. 2.4.

There is also evidence that when the ratio of the two cations on B sites is 1:3, long-range order can be established. Every row of octahedral cations in the $\langle 110 \rangle$ directions contains the B cations in ordered fashion, Fig. 2.9. An example is $(Fe)[Li_{0.5}Fe_{2.5}]O_4$ (Braun, 1952); the structure remains essentially cubic. The transition to a random distribution of Li^+ and Fe^{3+} on the octahedral B sites occurs between 1008 and 1028 K. A third type of cation ordering can occur when there is a 1:1 ratio of cations on the tetrahedral sites, in which the cations alternate on these sites. This superstructure has been observed in $Li_{0.5}Fe_{0.5}[Cr_2]O_4$ (Braun, 1957).

2.1.3   *Spinel solid solutions*

In addition to the extremely wide variety of simple binary spinels, it is possible to prepare many solid solutions series. The major advantage of forming solid solutions is that their physical properties vary continuously with composition, thus leading to the possibility of 'material design' for specific applications. The cations forming spinel solid solutions appear in Table 2.4. Solid solutions involving the substitution of one divalent cation by another, or by a combination of divalent cations, are first discussed.

A classical, extensively studied example is the so-called Ni–Zn ferrites, with the general formula:

$$Zn_xNi_{1-x}Fe_2O_4 \qquad (2.6)$$

where $0 \le x \le 1$. $NiFe_2O_4$ is an inverse spinel, and $ZnFe_2O_4$ is normal; the cation distribution is therefore:

$$(Zn_xFe_{1-x})[Ni_{1-x}Fe_{1+x}]O_4 \qquad (2.7)$$

These ferrites are the basis for many applications. The lattice parameter varies linearly with composition, x, as shown in Fig. 2.10. The cation radius ratio for tetrahedral sites is large (in four-fold symmetry, the Zn^{2+} radius is 0.60 Å, while that of Fe^{3+} is 0.49 Å (Shannon, 1976)), and has the strongest influence on the lattice parameter, as well as on the magnetic properties of these solid solutions (Globus, Pascard & Cagan, 1977).

Substitutions of Fe^{3+} have also been extensively studied. Blasse (1964) has reported the cation distribution in $Me^{2+}Fe_{2-x}Al_xO_4$, with $Me^{2+} =$ Fe, Co, Mg, Cu, Ni. The amount of Al on tetrahedral sites, y, is plotted as a function of x on Fig. 2.11, to show the cation distribution. In the case of $Fe^{2+}Fe_{2-x}^{3+}Al_xO_4$, for instance, the cation distribution is:

$$(Al_xFe_{1-x}^{3+})[Fe^{2+}Fe^{3+}]O_4 \qquad 0 \le x \le 1 \qquad (2.8)$$

$$(Al)[Fe^{2+}Al_{x-1}Fe_{2-x}^{3+}]O_4 \qquad 1 \le x \le 2 \qquad (2.9)$$

The substitution of Fe^{3+} by Cr^{3+}, Al^{3+} and Ti^{4+} has been studied by Gorter (1954).

The spinel systems containing Mn offer an additional complication: the energy difference between Mn^{2+} and Mn^{3+} is very small, and the following equilibrium exists:

$$Fe^{3+} + Mn^{2+} \rightleftarrows Fe^{2+} + Mn^{3+} \qquad 0.3\,eV \qquad (2.10)$$

Table 2.4. *Cations forming spinel solid solutions.*

	Valence		
1+	2+	3+	4+
Li	Mg	Al	Ti
Cu	Ca	Ti	V
Ag	Mn	V	Mn
	Fe	Cr	Ge
	Co	Mn	Sn
	Ni	Fe	
	Cu	Ga	
	Zn	Rh	
	Cd	In	

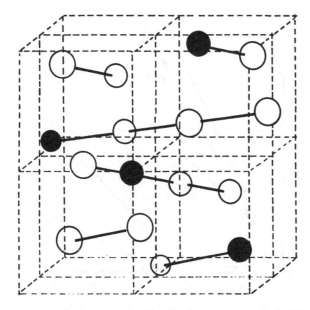

Fig. 2.9. Long-range order in spinels for 1:3 ratio of cations on octahedral sites (Aburto *et al.*, 1982).

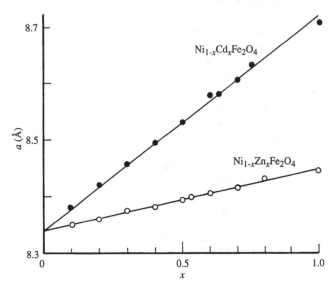

Fig. 2.10. Lattice parameters of $Zn_x Ni_{1-x} Fe_2 O_4$ and $Cd_x Ni_{1-x} Fe_2 O_4$ solid solutions as a function of composition x (Globus, Pascard & Cagan, 1977).

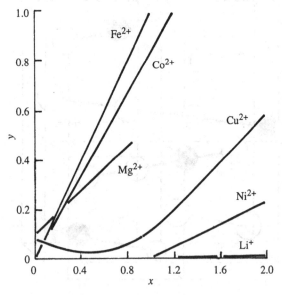

Fig. 2.11. Fraction of aluminium on tetrahedral sites, y, for spinel solid solutions $Me^{2+} Fe_{2-x} Al_x O_4$, with Me^{2+} = Fe, Co, Mg, Cu and Ni. (Adapted from Blasse, 1964.)

The tendency is generally to the left. Under oxidising conditions, however, some Mn^{2+} can be oxidised to produce an excess of Mn^{3+}, creating cation vacancies. In spinels, evidence of vacancies has been observed only in octahedral sites. The cation distribution in spinels containing simultaneously divalent and trivalent Fe and Mn has been studied by Rieck & Driessens (1966).

In most cases, the direct determination of cation distribution in spinels cannot be reliably established by x-ray diffraction, because the x-ray scattering factors for most of the transition-metal cations are very similar. Neutron diffraction experiments are more conclusive, since the scattering factors of transition metals for neutrons are quite different. Moreover, neutron spin interacts with crystal magnetic moments allowing a direct determination of the magnetic structure.

Neutron irradiation can produce drastic changes in the cation distribution. For radiation doses below 10^{18} cm^{-2}, no noticeable structural changes were detected in $Zn_{0.68}Ni_{0.32}Fe_2O_4$; but after a dose of 10^{20} cm^{-2}, the cation distribution was (Chukalkin *et al.*, 1975):

$$(Zn_{0.33}Fe_{0.67})[Ni_{0.32}Zn_{0.35}Fe_{1.33}]O_4 \qquad (2.11)$$

Mössbauer spectroscopy (recoiless absorption and reemission of γ radiation by ^{57}Fe nuclei) has been used to study the cation distribution in spinels (Watanabe *et al.*, 1982; Quintanar *et al.*, 1986; De Guire *et al.*, 1989). Mössbauer spectroscopy can provide detailed information about the Fe nucleus, such as its oxidation state, the symmetry of the crystal site it occupies and the magnitude of the local magnetic field.

A novel technique to determine the cation distribution in spinels involves the use of EXAFS (extended x-ray absorption fine structure) which seems to overcome the disadvantages of x-ray diffraction (Yao, Imafuji & Jinno, 1991).

2.1.4 Distorted spinels, magnetite and maghemite

In this section, unusual ferrite structures are briefly reviewed. Most of these ferrites show a non-cubic structure at low temperatures and the spinel structure at high temperatures. Examples of distorted spinels with $c/a > 1$ (tetragonal) are Mn_3O_4, $ZnMn_2O_4$, $NiCr_2O_4$ and $CuFe_2O_4$; $CuCr_2O_4$ shows the opposite distortion, $c/a < 1$. Magnetite, Fe_3O_4, also shows a distortion from cubic symmetry below 120 K, but its nature is associated with a complex ordering phenomenon. Finally, maghemite,

γ-Fe_2O_3, has a 'defective' spinel structure with a fraction of octahedral sites occupied by vacancies, at all temperatures.

With the exception of magnetite and maghemite, all these distortions can be explained in terms of the *Jahn–Teller* effect (Dunitz & Orgel, 1957a). For d^4 and d^9 transition-metal ions, a distortion from the regular octahedron can provide an additional stabilisation energy. For d^4 cations, and following Hund's rules, the lower triplet and the d_{z^2} orbitals are occupied by unpaired electrons; a simple mechanism to reduce electrostatic repulsion in the z-direction is to elongate the octahedron along the z-axis, Fig. 2.12. The structure becomes tetragonal with $c/a > 1$; this is the case

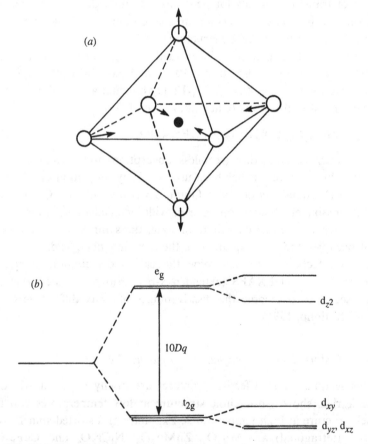

Fig. 2.12. (*a*) Deformation of a regular octahedron along the z-axis to produce $c/a > 1$. A small contraction is also produced in the x- and y-axes; (*b*) schematic energy level diagram showing the free-atom degenerated levels, the splitting for a regular octahedron and for a distorted octahedron. (Adapted from Dunitz & Orgel, 1957.)

for Mn_3O_4 and $ZnMn_2O_4$ (Mn^{3+} in octahedral sites). Cu^{2+} is a d^9 cation with all the orbitals full except the d_{z^2} orbital, where there is only one electron; this configuration leads to effects similar to those of d^4 cations. The energy levels are also shown in Fig. 2.12; the triplet is also split, since the elongation in the z-direction is accompanied by a small contraction of the x–y plane, thereby producing an energy difference between the d_{xy}, and the d_{zx} and d_{yz} orbitals.

Tetrahedra can also be distorted by the presence of d^8 and d^9 cations to gain additional stabilisation energy. A $c/a > 1$ distortion, Fig. 2.13, produces an additional splitting of the triplet and the doublet in such a way that the $d_{x^2-y^2}$ orbital becomes more stable, as well as the d_{xy} orbital in the triplet. Conversely, a $c/a < 1$ deformation leads to an additional stabilisation for d_{z^2} in the doublet, and d_{xz} and d_{yz} in the triplet, Fig. 2.14.

Nickel chromite, $NiCr_2O_4$, is a normal spinel (Romeijn, 1953) because of the high stabilisation energy of Cr^{3+} in octahedral sites; divalent Ni^{2+} is a d^8 cation which in tetrahedral coordination gains additional stabilisation energy with a $c/a > 1$ deformation. In the high-spin state, $d_{x^2-y^2}$, d_{z^2} and d_{xy} are full (paired electrons), while the orbitals d_{yz} and d_{xz} are half-filled. Copper chromite, $CuCr_2O_4$, is a normal spinel also due to the strong preference of Cr^{3+} for octahedral sites. Cu^{2+} (d^9) on tetrahedral sites is further stabilised by a small c/a distortion of the order of 0.91 (Miyahara & Ohnishi, 1956), as shown in Fig. 2.14.

A model for the transition from the low-temperature, distorted structure to the high-temperature, cubic structure has been proposed by Wojtowicz

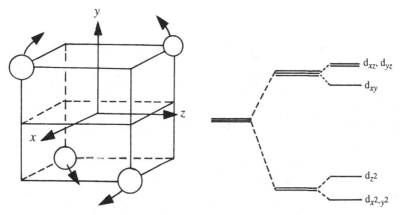

Fig. 2.13. Deformation of a tetrahedron to produce $c/a > 1$ and the corresponding energy level diagram. (Adapted from Dunitz & Orgel, 1957.)

(1959). According to this model, the transition is thermodynamically of the first order; i.e. latent heat, volume and lattice parameter discontinuities are expected, and have been observed in some of these compounds.

Magnetite, Fe_3O_4, is a unique material. It is a mixed-valency compound (Fe^{3+} and Fe^{2+} on crystallographic sites of the same symmetry) with a low electrical conductivity below ~ 120 K, and a nearly metallic conductivity above this temperature. Between 120 and ~ 770 K, magnetite is an inverse spinel, $(Fe^{3+})[Fe^{2+}Fe^{3+}]O_4$, and all the transition phenomena occur in the octahedral sites. For $T > 770$ K, the redistribution of Fe^{2+} on tetrahedral sites becomes non-negligible (Wu & Mason, 1981). The reversible, sharp discontinuity in conductivity was first observed by Okamura (1931). Ferroelectric features have also been observed at low temperatures (Rado & Ferrari, 1975; Kato *et al.*, 1983).

Verwey (1939) proposed that at low temperatures the unit cell remained cubic, with Fe^{2+} and Fe^{3+} ordered in alternate $\langle 110 \rangle$ directions; above the critical temperature, known as the *Verwey* transition, thermal energy allowed the transfer of electrons from Fe^{2+} to neighbouring Fe^{3+} cations, leading to high electrical ('hopping') conductivity. Fe^{2+} and Fe^{3+} become therefore indistinguishable. However, careful electron and neutron diffraction studies of synthetic crystals at low temperatures (Yamada, Suzuki & Chikazumi, 1968; Shirane *et al.*, 1975) showed evidence for a non-cubic unit cell. Mössbauer spectroscopy (Umemura & Iida, 1979) and nuclear magnetic resonance (NMR) results (Mizoguchi, 1978) showed that both Fe^{2+} and Fe^{3+} cations have several, different

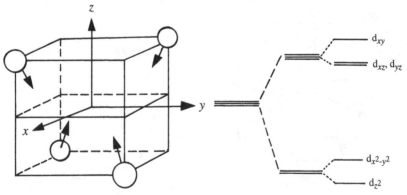

Fig. 2.14. Distorted tetrahedron with $c/a < 1$, and the corresponding energy level diagram. (Adapted from Dunitz & Orgel, 1957.)

environments, which is incompatible with a simple cubic symmetry. Iida *et al.* (1982) proposed a complex monoclinic unit cell for magnetite at low temperature.

The Verwey transition as well as the order of the phase change, is strongly dependent on the O/metal ratio. An increase in O content shifted the transition temperature from 120 to 81 K, with a variation in peak width from 2 to 11 K (Kakol, 1990). For $0 \leq x \leq 0.0117$ in the formula $Fe_{3-x}O_4$ (the equivalence between this notation and the δ term used in the original papers is $x = 3\delta$), a first-order transition was observed; for $0.0117 < x \leq 0.1053$, the transition showed a second- or higher-order character (Shepherd *et al.*, 1986). The value $x = 0.0117$ corresponded to the critical value at which one out of 32 ferrous cations (one in every two unit cells) was oxidised to a ferric ion. A similar trend and composition limits have been observed for $Fe_{3-x}Zn_xO_4$.

A simple model for the Verwey transition has been proposed (Honig, Spałek & Gopalan, 1990); octahedral sites in magnetite were represented by a site pair, with a ground energy state (an electron trapped), a first excited state (the electron resonating between the two components of the site pair) and a second excited state (two electrons in the site pair). An important characteristic of this model was that the Verwey transition was driven by the coulomb repulsive interaction between electrons in the site pair.

The surfaces of a magnetite single crystal have been studied by means of the scanning tunnelling microscope (STM). In this apparatus, a conducting probe is brought close to the surface to be studied; the probe is an extremely sharp sensor tip. An electron in the probe is attracted to the positive ions in the surface, and it can cross the empty space between the probe and the surface by a quantum-mechanical phenomenon known as *tunnelling*. Since tunnelling is extremely sensitive to the separation between the tip and the surface, monitoring the current as the tip is scanned near the surface leads to an accurate method for determining its topography. The resolution can be as small as ~ 0.01 Å (Freedman & Hansma, 1989). The *nanotopography* (surface features at the atomic scale) of magnetite was studied with non-magnetic as well as with magnetic sensor tips (Wiesendanger *et al.*, 1992). A contrast with a well-defined periodicity (12 Å) was observed in some regions with the magnetic tip, which was presumably due to Fe^{2+}–Fe^{3+} ordered configurations, in agreement with the model proposed by Iida *et al.* (1982). This was the first study in which magnetic imaging with the STM was observed.

2.2   Garnets

2.2.1   The garnet structure

The garnet ferrites are the basis of materials for many high-technology devices for magnetooptical, microwave and memory applications. These ferrites have thus been extensively studied as ceramics, single crystals, thin and epitaxial films, etc. The prototype, yttrium iron garnet or 'YIG' has also been used for many fundamental studies.

The crystal structure is that of the garnet mineral, $Mn_3Al_2Si_3O_{12}$. Si and Mn can be substituted by Y and Al by a double substitution mechanism:

$$3Mn^{2+} + 3Si^{4+} \leftrightarrows 3Y^{3+} + 3Al^{3+} \tag{2.12}$$

to obtain $Y_3Al_5O_{12}$. The magnetic garnets have Fe^{3+} instead of Al^{3+} and were prepared for the first time by Bertaut & Forrat (1956), and studied in detail by Geller & Gilleo (1957).

The general formula of ferrite garnets is $R_3Fe_5O_{12}$, where R is a rare-earth trivalent cation, or Y. The crystal structure has cubic symmetry and is relatively complex. The unit cell is formed by eight formula units (160 atoms) and belongs to the space group $O_h^{10} - Ia3d$. In contrast to spinels, the O sublattice is not a close-packed arrangement, and is better described as a polyhedra combination, Fig. 2.15.

The O polyhedra define three kinds of cation site: dodecahedral (eight-fold), octahedral (six-fold) and tetrahedral (four-fold). Rare earths, R, occupy the largest, dodecahedral sites, while Fe cations enter the octahedral and tetrahedral sites. The notation for the site occupancy is as follows:

$$\{R_3\}[Fe_2](Fe_3)O_{12} \tag{2.13}$$

{ } denotes dodecahedral sites, also known as c sites. There are 24 c sites in the unit cell; [ ] denotes octahedral sites, also designated as a sites, with 16 a sites per unit cell; and ( ) means tetrahedral sites or d sites; there are 24 d sites in each unit cell. An octant of the garnet unit cell is shown in Fig. 2.16.

All the polyhedra in garnets are distorted and twisted (Geschwind, 1961). Octahedral cations form a body centred cubic sublattice; the octahedra are distorted along one of the three-fold axes; this trigonal axis coincides with the [111] direction of the unit cell, Fig. 2.17. The octahedra are rotated about the [111] direction through opposite angles of ±28°. Octahedron edges in YIG are 2.68 and 2.99 Å (Geller & Gilleo, 1957).

The tetrahedra are also distorted. If the tetrahedron is inscribed within a cube whose axes are parallel to the unit cell, the distortion is along the unit cell edge, and the cube is rotated through angles of ±16°, Fig. 2.18. Two edge lengths have been observed in tetrahedra in YIG: 3.16 and 2.87 Å. The dodecahedra are cubes with faces slightly deformed along the cube diagonal, Fig. 2.19, with four different cube edge lengths; 2.68, 2.87 and 2.96 Å.

Each octahedron shares six edges with dodecahedra and only corners with tetrahedra. Each tetrahedron shares two edges with dodecahedra: octahedra and dodecahedra share only corners. Each dodecahedron

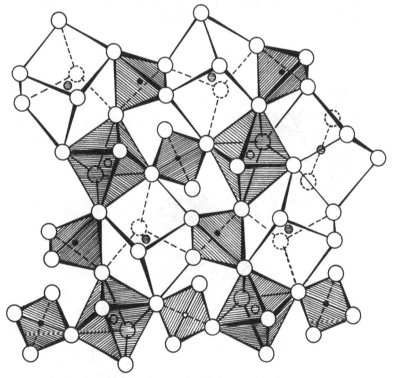

Fig. 2.15. Garnet structure described as a polyhedra combination. Large circles are O atoms, small full circles are tetrahedral sites, small empty circles are octahedral sites and shaded circles are dodecahedral sites. (Adapted from Gilleo, 1980.)

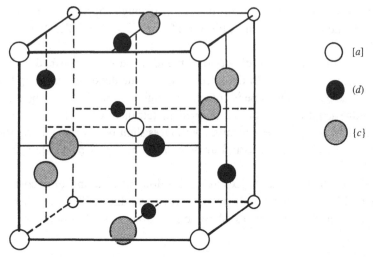

Fig. 2.16. An octant of the garnet unit cell showing cations on octahedral, tetrahedral and dodecahedral sites. O atoms have been omitted for clarity. (Adapted from Geller & Gilleo, 1957.)

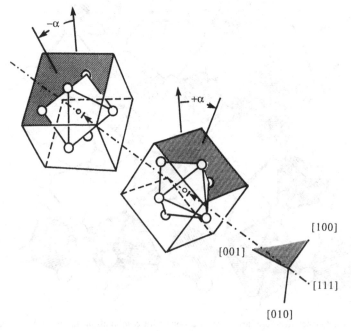

Fig. 2.17. Octahedral sites in garnets, showing the distortion along the ⟨111⟩ directions. The angle α is $\pm 28°$ (Geschwind, 1961).

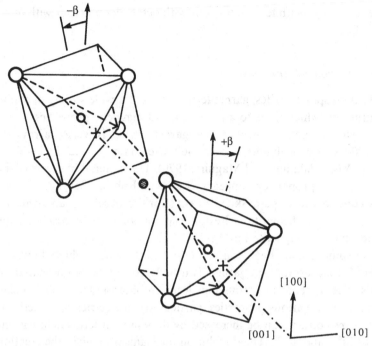

Fig. 2.18. Tetrahedral site distortions in garnets along the ⟨110⟩ directions. The angle β is ±16° (Geschwind, 1961).

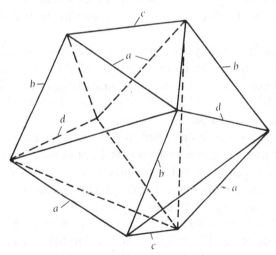

Fig. 2.19. Dodecahedral polyhedron in *c* sites; edge lengths are (in Å): *a* 2.68; *b* 2.81; *c* 2.87; and *d* 2.96. (Adapted from Geller & Gilleo, 1957.)

shares two edges with tetrahedra, four with octahedra and four with other
dodecahedra.

2.2.2   Cation substitutions

Similarly to spinel ferrites, garnet ferrites present a wide variety of cation
substitutions, which leads to a large range of magnetic properties.

The stoichiometry range of the garnet phase is very narrow; at
$T < 1000\,°C$ it is negligible, and at $1460\,°C$ contains ~ 37.2–37.5 mole $\%$
Y_2O_3 in YIG (Paladino and Maguire, 1970). The material in excess of the
stoichiometric proportions appears as a second phase. For Y_2O_3 excess,
the second phase is perovskite, $YFeO_3$; for Fe excess in an oxidising
atmosphere, it is hematite, α-Fe_2O_3, and for an inert atmosphere, the
second phase is magnetite Fe_3O_4.

As in spinels, some Fe^{2+} can be present in the garnet phase; however,
the Fe^{2+} concentration in garnets is usually considerably lower than in
spinels. The presence of ferrous ions leads to photo-induced phenomena
and changes in various properties, particularly in electrical conductivity.
The ferrous content can be increased by doping with tetravalent cations
such as Si^{4+} and Ti^{4+}. The dissolution mechanism involves the creation
of point defects (Metselaar & Huyberts, 1973). Ca^{2+} doping of YIG
produces a p-type semiconductor, with an electrical conductivity several
orders of magnitude higher than undoped YIG, by oxidation of some of
the Fe cations to Fe^{4+} (Thravendrarajah, Pardavi-Horvat & Wigen,
1990).

Rare-earth cations usually enter the large dodecahedral sites due simply
to their cation radius. The lattice constants of rare-earth garnets are
presented in Table 2.5. In addition to the rare earths listed in Table 2.5,
Nd^{3+}, Pr^{3+}, Ce^{3+}, Bi^{3+}, Ca^{2+}, La^{3+}, Mn^{2+}, Na^+, Sr^{2+}, Pb^{2+} and Fe^{2+}
enter c sites, at least as a partial substitution.

The use of new preparation techniques (see next chapter) has led, in
some cases, to non-equilibrium compounds. For example, BiIG (bismuth
iron garnet, $Bi_3Fe_5O_{12}$), which has important magnetooptical properties
(see Section 4.8), had not been successfully prepared in the full-substituted
form by conventional methods, but only as $Y_{3-x}Bi_xFe_5O_{12}$ with $x \leq 1.7$
(Hansen, Witter & Tolksdorf, 1983); by using rf-magnetron sputtering,
however, Okuda *et al.* (1990) have obtained thin films of BiIG with a
lattice constant of 12.631 Å, the largest known in ferrite garnets.

Some cations show a strong preference for octahedral sites due to the

Table 2.5. *Lattice constants of
rare-earth substituted garnets.*

Formula	Lattice constant (Å)
$Lu_3Fe_5O_{12}$	12.283
$Yb_3Fe_5O_{12}$	12.302
$Tm_3Fe_5O_{12}$	12.323
$Er_3Fe_5O_{12}$	12.347
$Ho_3Fe_5O_{12}$	12.375
$Dy_3Fe_5O_{12}$	12.405
$Tb_3Fe_5O_{12}$	12.436
$Gd_3Fe_5O_{12}$	12.471
$Eu_3Fe_5O_{12}$	12.498
$Sm_3Fe_5O_{12}$	12.529
$Y_3Fe_5O_{12}$	12.376

From Geller & Gilleo (1957), Espinosa
(1962) and Geller, Williams & Sherwood
(1961).

contribution of crystal field stabilisation to the lattice energy. The final
cation distribution is also influenced by the thermal history of the sample.

Cr^{3+}, Sc^{3+} and In^{3+} show a strong octahedral preference (Gilleo,
1980). Also found to occupy octahedral sites are: Mn^{2+}, Mn^{4+}, Co^{2+},
Cu^{2+}, Mn^{3+}, Fe^{3+}, Fe^{2+}, Ni^{2+}, Ru^{3+}, Ru^{4+}, Ir^{4+}, Al^3, Ge^{4+}, Ti^{4+},
Co^{3+}, Sb^{5+}, Ga^{3+}, Sn^{4+}, Hf^+, Mg^{2+} and Zr^{4+}. In some cases, rare-earth
ions also enter octahedral sites; for example, in the case of the gallates
$\{Nd_3\}[R_2](Ga_3)O_{12}$ where R = Lu, Yb, Tm, Er, Ho and Dy, a complete
substitution has been observed (Suchow & Kokta, 1972). Tetrahedral sites
can be occupied by the following cations: Ga^{3+}, Al^{3+}, Mn^{2+}, Co^{3+}, Ni^{2+},
Si^{4+}, V^{5+}, Ge^{4+}, Ti^{4+}, Sn^{4+} and Fe^{4+}. As can be seen, many cations
can occupy both octahedral and tetrahedral sites.

2.2.3 Garnet solid solutions

The garnet structure also forms a wide range of complete solid solutions.
Rare-earth ions can be substituted partially or totally on *c* sites; examples
of these solutions are $Y_{3-x}Sm_xFe_5O_{12}$ and $Gd_{3-x}Eu_xFe_5O_{12}$ (Heilner &
Grodkiewicz, 1973) and $Y_{3-x}Lu_xFe_5O_{12}$ (Gyorgy *et al.*, 1973). In all these
cases, lattice parameters vary linearly with composition.

Solid solutions of cations occupying both a and d sites often lead to redistribution phenomena on thermal treatment. Cases extensively studied are $Y_3Fe_{5-x}Ga_xO_{12}$ and $Y_3Fe_{5-x}Ga_xO_{12}$. Ga^{3+} and Al^{3+} have some preference for tetrahedral sites which can be altered by annealing and quenching. The preference of Ga^{3+} for d sites is stronger than that of Al^{3+}. The activation energy for the redistribution process has been calculated by several authors, with values ranging from 1 to 1.3 eV for Ga (Kurtzig & Dixon, 1972) and 1 to 1.7 eV for Al (Röschmann & Hansen, 1981). The influence of the rare-earth cation on c sites is important, as is shown by a redistribution activation energy of 0.3 eV for $Ho_3Fe_{5-x}Al_xO_{12}$ (Ostoréro *et al.*, 1985).

For relatively low-temperature annealing, $T \leq 900\,°C$, the time needed to achieve the equilibrium distribution is long, and can be drastically lowered by using an inert or slightly reducing atmosphere (Kurtzig & Dixon, 1972). The formation of O vacancies enhances diffusion. The changes in the Fe/Ga distribution produced by quenching are reversible for $900 \leq T \leq 1250\,°C$ (Enoch *et al.*, 1976).

2.3   Hexagonal ferrites

2.3.1   Crystal structures

Rather than a single structure type, hexagonal ferrites are a numerous *family* of related compounds with hexagonal and rhombohedral symmetry. All of them are synthetic except magnetoplumbite, of approximate formula $PbFe_{7.5}Mn_{3.5}Al_{0.5}Ti_{0.5}O_{19}$, which is the only natural compound isomorphous with barium ferrite, $BaFe_{12}O_{19}$.

The composition of the various compounds can be understood by examining the upper section of the ternary phase diagram $MeO–Fe_2O_3–BaO$, where Me is a divalent metal such as Ni, Mg, Co, Fe, Zn, Mn or Cu, Fig. 2.20. All the ferrites are found on the joins $BaFe_{12}O_{19}–Me_2Fe_4O_8$ and $BaFe_{12}O_{19}–Me_2BaFe_{12}O_{22}$, or M–S and M–Y, respectively. M, S and Y are the end-members.

On the join M–S, the compounds MS (or W), M_2S (or X), M_4S and M_6S have been synthesised; on the join M–Y, up to 40 compounds in the series M_2Y_n are known, and 11 compounds in the series M_4Y_n have been prepared. This remarkable variety of compounds is the result of a building principle based on block unit stacking, which will now be discussed.

The basic block units are denoted as S, R and T blocks; all of them have a close-packed arrangement of O atoms. The S unit is formed by two formula units $MeFe_2O_4$ with the spinel structure thus containing two tetrahedral and four octahedral cation sites. The R block has the stoichiometry $(BaFe_6O_{11})^{2-}$, with three octahedral sites of two different types and one bipyramidal, five-fold cation site; the Ba^{2+} occupies an O^{2-}, 12-fold site, Fig. 2.21. The T unit has four layers with formula $Ba_2Fe_8O_{14}$, where Fe ions occupy two octahedral sites, and six octahedral sites of two different types. The block lengths are 4.81 Å, 11.61 Å and 14.52 Å for the S, RS and TS blocks, respectively (Winkler, 1971).

The unit cell of the M structure contains two formula units, $BaFe_{12}O_{19}$, and includes the unit block RSR*S*, where the asterisks indicate a 180° rotation with respect to the c-axis, Fig. 2.22. The unit cell parameters are $a = 5.892$ Å and $c = 23.183$ Å (Obradors *et al.*, 1985). The RS blocks in this case are formed by the group $(BaFe_6O_{11})^{2-}$; the S block is $(Fe_6O_8)^{2+}$ and has no Me cations. The close-packing arrangement of

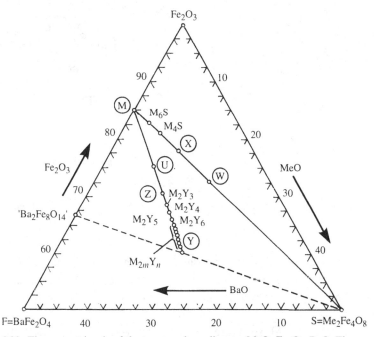

Fig. 2.20. The upper triangle of the ternary phase diagram MeO–Fe_2O_3–BaO. The composition of end-members and important compounds is indicated. (Adapted from Winkler, 1971.)

Table 2.6. *The five cation sublattices in the M hexaferrites.*

Sublattice	Number of sites	Symmetry	Block
k	12	octahedral	R, S
f_{vi}	4	octahedral	R
a	2	octahedral	S
f_{iv}	4	tetrahedral	S
b	2	five-fold	R

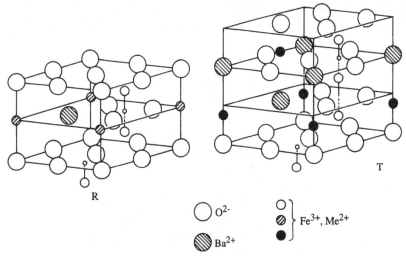

Fig. 2.21. Structure of the R and T block units in hexagonal ferrites. (Adapted from Smit & Wijn, 1961.)

oxygen has cubic symmetry in the S units and hexagonal symmetry in the R blocks. The (111) spinel axis coincides with the hexagonal axis. The crystallographic space group of the M structure is $P6_3/mmc$.

Fe ions in the M structure occupy five types of crystal site: three non-equivalent octahedral sites, tetrahedral sites and bipyramidal, five-fold sites, summarised in Table 2.6. There are 12 octahedral, k sites, four

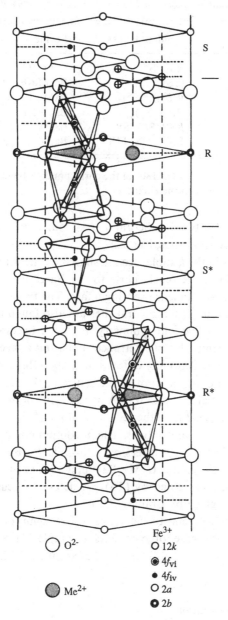

S

R

S*

R*

O O^{2-}

Fe^{3+}

$\circ$ $12k$

$\circledcirc$ $4f_{vi}$

$\bullet$ $4f_{iv}$

$\bigcirc$ Me^{2+}

$\circ$ $2a$

$\bullet$ $2b$

Fig. 2.22. Unit cell of $BaFe_{12}O_{19}$, showing the polyhedra coordination for Fe in f_{iv} and f_{vi} sites. The common faces of two neighbouring f_{iv} and f_{vi} polyhedra are hatched (Albanese, Carbucicchio & Deriu, 1973).

Table 2.7. *Building blocks and formulae of hexagonal ferrites.*

Designation	Building blocks	Formula
M	RSR*S*	$BaFe_{12}O_{19}$
Y	$(TS)_3$	$Ba_2Me_2Fe_{12}O_{22}$
W (MS)	$R(S)_2R*(S*)_2$	$BaMe_2Fe_{16}O_{27}$
X (M_2S)	$(RSR*S_2^*)_3$	$Ba_2Me_2Fe_{28}O_{46}$
U (M_2Y)	RSR*S*T*S*	$Ba_4Me_2Fe_{36}O_{60}$
Z (M_2Y_2)	RSTSR*S*T*S*	$Ba_6Me_4Fe_{48}O_{82}$

* Indicates a rotation of $180°$ of the corresponding building unit with respect to the c-axis.

octahedral, f_{vi} (or f_2), two octahedral a sites, four tetrahedral f_{iv} (or f_1) sites, and two b, five-fold sites, Fig. 2.22.

The 12 iron sites in the k sublattice are shared by the R and the S units. The four f_{vi} are in the R block, close to the Ba cation and forming a (Fe_2O_9) group of two octahedra sharing a face. The two a sites are found in the S block; the four f_{iv} tetrahedral sites are also found in the S block. The two b sites with five-fold symmetry are quite uncommon in ferric oxides. They are formed by two tetrahedra sharing a face, occupied by only one cation. A detailed x-ray diffraction study (Obradors *et al.*, 1985) showed that these Fe ions are not in the centre of the bipyramid, but in a double-well potential of width 0.17 Å at each side of the centre.

As indicated in Table 2.7, all the related compounds can be understood as a combination of the unit blocks. The Y compound, for instance, has a T block which has four O layers with formula $Ba_2Fe_8O_{14}$ and an S block with $Me_2Fe_4O_8$, leading to $Me_2Ba_2Fe_{12}O_{22}$. The unit cell has three times these atoms, represented by $(TS)_3$. A detailed discussion of these structures was first given by Braun (1957).

Hexagonal ferrites seem to be quite stable with deviations from stoichiometry; in the case of commercial samples of $SrFe_{12}O_{19}$, for example, a single phase was observed for Fe–Sr ratios of 12.7:1 (Thomson & Evans, 1990).

2.3.2   Cation substitutions

Hexagonal ferrites also show a wide range of cation substitutions. In many of them, Ba can be substituted by Sr, Ca and Pb. The substitution of Ba

Table 2.8. *Cation substitutions in barium hexaferrite* M $Ba_{1-x}A_x^{k+}Fe_{12-y}B_y^{l+}O_{19}$.

A			B			
$k = 1$	2	3	$l = 2$	3	4	5
Na	Sr	La	Mg	Cr	Ti	Sb
K	Pb	Pr	Mn	Mn	Ir	As
Rb	Ca	Nd	Fe	Co	Ge	V
Ag		Sm	Co	Al	Sn	Ta
Tl		Eu	Ni	Ga	Zr	Nb
		Bi	Zn	In		
			Cu	Sb		
				Sc		
				Ru		

Adapted from Kools, 1991.

by Sr in M ferrites is important technologically; as Sr^{2+} is smaller than Ba^{2+}, the magnetic ions are closer and establish stronger interactions. Some La^{3+} can substitute for Ba^{2+}, but an equivalent amount of Fe^{3+} reduces to Fe^{2+} to preserve electrical neutrality (Smit & Wijn, 1961). There is also some evidence that Ce^{3+} and Ce^{4+} can substitute for Ba^{2+} in W compounds (Kui, Lu & Du, 1983), and Gd^{3+} for Ba^{2+} in X compounds (Gu, Lu & Du, 1983), presumably also by an equivalent reduction of Fe^{3+} to Fe^{2+} ions for charge compensation.

Fe^{3+} can be replaced by trivalent cations such as Al^{3+}, Ga^{3+}, In^{3+}, Sc^{3+} (Albanese *et al.*, 1980), or a divalent + tetravalent combination such as $Co^{2+} + Ti^{4+}$ in $BaCo_xTi_xFe_{12-2x}O_{19}$ (Casimir *et al.*, 1959), or $Mn^{2+} + Ti^{4+}$ in $BaMn_xTi_xFe_{12-2x}O_{19}$ (Turilli, Licci & Rinaldi, 1986). A summary of reported cation substitutions in M ferrites is shown in Table 2.8.

In W, U, X, Y, and Z compounds, the Me divalent cation can be Co, Zn, Ni, Fe, Cu, Mg or Mn, or a combination of them. The resulting cation distributions are, of course, very complex, since in many cases, cations of a given kind can occupy more than one site type. In the BaCo–W compound ($BaCo_2Fe_{16}O_{27}$), for example, neutron diffraction studies (Samaras *et al.*, 1989) showed that Co^{2+} occupied four of the seven non-equivalent cation sublattices of this structure.

2.4 Microstructure

2.4.1 The polycrystalline state

The melting point of the magnetic ceramics is usually very high; for many compounds, an irreversible O loss takes place before melting occurs. The preparation of single crystals by simple melting and recrystallisation is not a feasible technique. In some cases, single crystals have been successfully prepared by other methods, but the most general (and economical) route to obtain practically any ceramic is sintering, which gives *polycrystals*. The preparation techniques are discussed in the next chapter; in this section, the nature and consequences of polycrystallinity will be briefly reviewed.

A polycrystal is much more than many tiny crystals bonded together. The interface between the crystals, or the *grain boundaries* which separate and bond the grains, are complex and interactive interfaces. The whole set of a given material's properties (mechanical, chemical and especially electrical and magnetic) depend strongly on the nature of the microstructure.

In the simplest case, the grain boundary is the region which accommodates the difference in crystallographic orientation between the neighbouring grains. For certain simple arrangements, the grain boundary is made of an array of dislocations, whose number and spacing depends on the angular deviation between the grains. The ionic nature of ceramics leads to dislocation patterns considerably more complex than in metals, since electrostatic energy accounts for a significant fraction of the total boundary energy. Some grain boundary structures in ceramics (NiO) have been reviewed by Fischmeister (1987).

Interfaces play a central role in sintering and any other diffusion phenomena; for instance, diffusion *along* grain boundaries can be ~100 times faster than *across* them (Kingery, 1984). Interfaces are a common location for segregated second phases; in fact, grain boundaries in ceramics are more prone to segregation than their counterpart in metals because segregation can be promoted by electrostatic interactions (Kingery, 1974). When segregation occurs, grain boundaries become 'inhomogeneous', as their chemical composition is different from that of the grains. Large deviations from stoichiometry have been observed in grain boundaries of spinels and titanates (Chiang & Peng, 1987).

A classical example of the advantages of a controlled segregation in ferrites is the addition of CaO to Mn spinels. The presence of CaO in the grain boundaries increases the electrical resistivity thus reducing eddy current losses when magnetic flux frequency variations occur at a high frequency (Pinet-Berger & Laval, 1987). The presence of larger foreign cations in the grain boundaries can also produce significant strains in neighbouring grains (Nan Lin, Mishra & Thomas, 1984).

The uniformity in the grain size and the average grain diameter can control properties such as the magnetic permeability. An undesirable effect of certain sintering conditions is the formation of the so-called 'duplex' microstructure, where a very large grain is surrounded by smaller ones, Fig. 2.23. When this occurs, the large grain has a high defect concentration.

Magnetic oxides have been obtained as an *amorphous* phase, with no

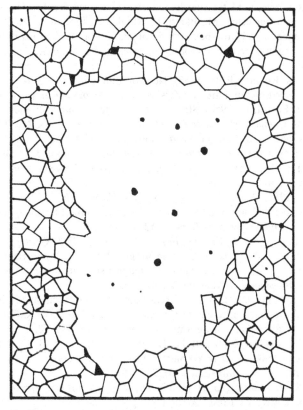

Fig. 2.23. A 'duplex' structure, characterised by a large, defective grain surrounded by smaller grains (schematic).

The crystal structures of magnetic ceramics

long-range atomic order. Since magnetic interactions tend to be weaker in disordered phases, most of amorphous ferrites are magnetic only at low temperatures. The microstructure in the systems $CaO-Bi_2O_3-Fe_2O_3$ and $Li_2O-Bi_2O_3-Fe_2O_3$ has been studied by electron microscopy (Nakamura, Hirotsu & Ichinose, 1991); other systems have been reviewed by Sugimoto (1989).

References

Aburto, S., Jiménez, M., Marquina, M. L. & Valenzuela, R. (1982). The molecular field approximation in Ni–Zn ferrites. In *Ferrites: Proceedings of the Third International Conference*. Eds: H. Watanabe, S. Iida and M. Sugimoto. Center for Academic Publications, Tokyo, pp. 188–91.

Albanese, G., Carbucicchio, M. & Deriu, A. (1973). Substitution of Fe^{3+} by Al^{3+} in the trigonal sites of M-type hexagonal ferrites. *Nuovo Cimento*, **15B**, 147–58.

Albanese, G., Carbucicchio, M., Pareti, L., Rinaldi, S., Licchini, E. & Slokar, G. (1980). Magnetic and Mössbauer study of Al, Ga, In and Sc-substituted Zn_2-W hexagonal ferrites. *Journal of Magnetism and Magnetic Materials*, **15–18**, 1453.

Bacon, G. E. & Roberts, F. F. (1953). Neutron diffraction studies of magnesium-ferrite-aluminate powders. *Acta Crystallographica*, **6**, 57–62.

Bertaut, F. & Forrat, F. (1956). Structure des ferrites ferrimagnétiques des terres rares. *Comptes Rendus de l'Academie des Sciences, Paris*, **242**, 328–84.

Blasse, G. (1964). Crystal chemistry and some magnetic properties of mixed oxides with spinel structure. *Philips Research Reports*, Suppl. 3, 1–139.

Bragg, W. H. (1915). The structure of the spinel group of crystals. *Philosophical Magazine*, **30**, 305–15.

Braun, P. B. (1952). A superstructure in spinels. *Nature*, **170**, 1123.

Braun, P. B. (1957). The crystal structure of a new group of ferromagnetic compounds. *Philips Research Reports*, **12**, 491–548.

Casimir, H. B. G., Smit, J., Enz, U., Fast, J. F., Wijn, H. P. J., Gorter, E. W., Duyvesteyn, A. J. W., Fast, J. D. & de Jong, J. J. (1959). Rapport sur quelques recherches dans le domaine du magnétisme aux laboratoires Philips. *Journal de Physique et le Radium*, **20**, 360–73.

Chiang, Y. M. & Peng, C-J. (1987). Grain-boundary non-stochiometry in spinels and titanates. *Advances in Ceramics*, Vol. 23. Eds C. R. A. Catlow and N. C. Mackrodt. American Ceramic Society, Ohio, pp. 361–78.

Chukalkin, Y. G., Goshchitskii, B. N., Dubinin, S. F., Sidorov, S. K., Petrov, Y. V., Parkhomenko, V. D. & Vologin, V. G. (1975). Radiation effects in oxide ferrimagnets. *Physica status solidi (a)*, **28**, 345–54.

Condon, E. U. & Shortley, G. H. (1935). *The Theory of Atomic Spectra*. Cambridge University Press, London.

Corliss, L. M., Hastings, J. M. & Brockman, F. G. (1953). A neutron diffraction study of magnesium ferrite. *Physical Review*, **90**, 1013–18.

References

Cracknell, A. P. (1975). *Magnetism in Crystalline Materials.* Pergamon Press, Oxford, pp. 138–57.

de Boer, F., van Santen, J. H. & Verwey, E. J. W. (1950). The electrostatic contribution to the lattice energy of some ordered spinels. *Journal of Chemical Physics*, **18**, 1032–4.

De Guire, M. R., O'Handley, R. C. & Kalonji, G. (1989). The cooling rate dependence of cation distributions in $CoFe_2O_4$. *Journal of Applied Physics*, **65**, 3167–72.

Dunitz, J. D. & Orgel, L. E. (1957). Electronic properties of transition metal oxides: Part II, Cation distribution amongst octahedral and tetrahedral sites. *Journal of Physics and Chemistry of Solids*, **3**, 318–23.

Dunitz, J. D. & Orgel, L. E. (1957a). Electronic properties of transition metal oxides: Part I, Distortions from cubic symmetry. *Journal of Physics and Chemistry of Solids*, **3**, 20–9.

Enoch, R. D., Jones, M. E., Murrell, D. L., Fiddyment, P. J. & Waters, D. G. P. (1976). Heat-treated behavior of the magnetic properties of epitaxial $Sm_{0.4}Y_{2.6}Fe_{3.8}Ga_{1.2}O_{12}$. *Journal of Applied Physics*, **47**, 2705–9.

Espinosa, G. P. (1962). Crystal chemical study of the rare earth iron garnets. *Journal of Chemical Physics*, **37**, 2344–7.

Fischmeister, H. F. (1987). Progress in the understanding of ceramic microstructures and interfaces since 1976. In *Ceramic Microstructures '86: Role of Interfaces*. Materials Science Research Series, Vol. 21. Eds J. A. Pask and A. G. Evans. Plenum Press, New York, pp. 1–14.

Freedman, R. A. & Hansma, P. K. (1989). The scanning tunneling microscope. In *Modern Physics*. Eds R. A. Serway, C. J. Moses and C. A. Moyer. Saunders Golden Sunburst Series, Saunders College Publishing, Philadelphia, pp. 187–93.

Fuentes, V., Aburto, S. & Valenzuela, R. (1987). Magnetic sublattices in nickel ferrite. *Journal of Magnetism and Magnetic Materials*, **69**, 233–6.

Geller, S. & Gilleo, M. A. (1957). The crystal structure and ferrimagnetism of yttrium-iron garnet $Y_3Fe_2(FeO_4)_3$. *Journal of Physics and Chemistry of Solids*, **3**, 30–6.

Geller, S., Williams, H. J. & Sherwood, R. C. (1961). Magnetic and crystallographic study of neodymium-substituted yttrium and gadolinium iron garnets. *Physical Review*, **123**, 1692–9.

Geschwind, S. (1961). Paramagnetic resonance of Fe^{3+} in octahedral and tetrahedral sites in yttrium-gallium garnet (YGaG) and anisotropy of yttrium-iron garnet (YIG). *Physical Review*, **121**, 363–74.

Gilleo, M. A. (1980). Ferromagnetic insulators: Garnets. In *Ferromagnetic Materials*, Vol. 2. Ed. E. R. Wohlfarth. North Holland, Amsterdam, pp. 1–54.

Globus, A., Pascard, H. & Cagan, V. (1977). Distance between magnetic ions and fundamental properties in ferrites. *Journal de Physique Colloque C1*, **C1–38**, C1-163–8.

Gorter, E. W. (1954). Saturation magnetization and crystal chemistry of ferrimagnetic oxides. *Philips Research Reports*, **9**, 295–320.

Gu, B., Lu, H. & Du, Y. (1983). Magnetic properties and Mössbauer spectra of

X-type hexagonal ferrites. *Journal of Magnetism and Magnetic Materials,* **31–34,** 803–4.

Gyorgy, E. M., Sturge, M. D., Van Uitert, L. G., Heilner, E. J. & Grodkiewicz, W. H. (1973). Growth-induced anisotropy of some mixed rare-earth iron garnets. *Journal of Applied Physics,* **44,** 438–43.

Hansen, P., Witter, K. & Tolksdorf, W. (1983). Magnetic and magneto-optic properties of lead- and bismuth-substituted yttrium iron garnet films. *Physical Review B,* **27,** 6608–25.

Hastings, J. M. & Corliss, L. M. (1953). Neutron diffraction studies of zinc ferrite and nickel ferrite. *Reviews of Modern Physics,* **25,** 114–19.

Hastings, J. M. & Corliss, L. M. (1956). Neutron diffraction study of manganese ferrite. *Physical Review,* **104,** 328–31.

Heilner, E. J. & Grodkiewicz, W. H. (1973). Compositional dependence of cubic and uniaxial anisotropies in some mixed rare-earth garnets. *Journal of Applied Physics,* **44,** 4218–19.

Hill, R. J., Craig, J. R. & Gibbs, G. V. (1979). Systematics of the spinel structure type. *Physical Chemistry of Minerals,* **4,** 317–39.

Honig, J. M., Spałek, J. & Gopalan, P. (1990). Simple interpretation of the Verwey transition in magnetite. *Journal of the American Ceramic Society,* **73,** 3225–30.

Iida, S., Mizushima, K., Mizoguchi, M., Kose, K., Kato, K., Yanai, K., Goto, N. & Yumoto, S. (1982). Details of the electronic superstructure of Fe_3O_4. *Journal of Applied Physics,* **53,** 2164–6.

Jagodzinski, H. (1970). Crystallographic aspects of non-stoichiometry of spinels. In *Problems of Nonstoichiometry.* Ed. A. Rabenau. North-Holland, Amsterdam, pp. 131–77.

Kakol, Z. (1990). Magnetic and transport properties of magnetite in the vicinity of the Verwey transition. *Journal of Solid State Chemistry,* **88,** 104–14.

Kato, K., Iida, S., Yanai, K. & Mizushima, K. (1983). Ferrimagnetic ferroelectricity of Fe_3O_4. *Journal of Magnetism and Magnetic Materials,* **31–34,** 783–4.

Kimura, T., Ichikawa, M. & Yamaguchi, T. (1977). Effects of grain size on cation ordering in sintered Mg-ferrites. *Journal of Applied Physics,* **48,** 5033–7.

Kingery, W. D. (1974). Plausible concepts necessary and sufficient for interpretation of ceramic grain-boundary phenomena: I, Grain boundary characteristics, structure and electrostatic potential; II, Solute segregation, grain-boundary diffusion and general discussion. *Journal of the American Ceramic Society,* **57,** 1–8 and 74–83.

Kingery, W. D. (1984). Segregation phenomena at surfaces and at grain boundaries in oxides and carbides. *Solid State Ionics,* **12,** 299–307.

Kools, F. (1991). Hard ferrites. In *Concise Encyclopedia of Advanced Ceramic Materials.* Ed. R. J. Brook. Pergamon Press, Oxford, pp. 200–6.

Kui, J., Lu, H. & Du, Y. (1983). W-type hexagonal ferrites $R_xBa_{1-x}Fe_{18}O_{27}$. *Journal of Magnetism and Magnetic Materials,* **31–34,** 801–2.

Kurtzig, A. J. & Dixon, M. (1972). Control of magnetization of bubble garnets by annealing. *Journal of Applied Physics,* **43,** 2883–5.

References

Lotgering, F. K. (1956). On the ferrimagnetism of some sulfides and oxides. *Philips Research Reports*, **11**, 190–249.

McClure, D. S. (1957). The distribution of transition metal cations in spinels. *Journal of Physics and Chemistry of Solids*, **3**, 311–17.

Metselaar, R. & Huyberts, M. A. H. (1973). The stoichiometry and defect structure of yttrium iron garnet and the nature of the centres active in the photomagnetic effect. *Journal of Physics and Chemistry of Solids*, **34**, 2257–63.

Miyahara, S. & Ohnishi, H. (1956). Cation arrangement and magnetic properties of copper ferrite-chromite series. *Journal of the Physical Society, Japan*, **11**, 1296–7.

Mizoguchi, M. (1978). NMR study of the low temperature phase of Fe_3O_4. I. Experiments; II. Electron ordering analysis. *Journal of the Physical Society of Japan*, **44**, 1501–20.

Nakamura, S., Hirotsu, Y. & Ichinose, N. (1991). High resolution electron microscopic observation of ferromagnetic amorphous ferrites in the $CaO–Bi_2O_3–Fe_2O_3$ and $Li_2O–Bi_2O_3–Fe_2O_3$ systems. *Japanese Journal of Applied Physics*, **30**, L844–7.

Nan Lin, I., Mishra, R. K. & Thomas, G. (1984). Interaction of magnetic domain walls with microstructural features in spinel ferrites. *IEEE Transactions on Magnetics*, **20**, 134–9.

Nishikawa, S. (1915). Structure of some crystals of the spinel group. *Proceedings of the Mathematical and Physical Society of Tokyo*, **8**, 199–209.

Obradors, X., Collomb, A., Pernet, M., Samaras, D. & Joubert, J. C. (1985). X-ray analysis of the structural and dynamical properties of $BaFe_{12}O_{19}$ hexagonal ferrite at room temperature. *Journal of Solid State Chemistry*, **56**, 171–81.

Okamura, T. (1931). Transformation of magnetite at low temperature. *Scientific Reports, Tohoku University*, **21**, 231–41.

Okuda, T., Katayama, T., Kobayashi, H., Kobayashi, N., Satoh, K. & Yamamoto, H. (1990). Magnetic properties of $Bi_3Fe_5O_{12}$ garnet. *Journal of Applied Physics*, **67**, 4944–6.

Ostoréro, J., Le Gall, H., Makram, H. & Escorne, M. (1985). The influence of quenching temperature on the magnetic properties of rare-earth garnet HoIG: Al single crystals. In *Advances in Ceramics: Fourth International Conference on Ferrites*, Vol. 15. Ed. F. F. Y. Wang. American Ceramic Society, Ohio, pp. 265–73.

Paladino, A. E. & Maguire, E. A. (1970). Microstructure development in yttrium-iron garnet. *Journal of the American Ceramic Society*, **53**, 98–102.

Pinet-Berger, J. J. & Laval, J. Y. (1987). The crystallographical and chemical relationship between intergranular and bulk resistivity in semiconductor oxides. In *Ceramic Microstructures: Role of Interfaces*, Vol. 21. Ed. J. A. Pask and A. G. Evans. Materials Science Research Series, Plenum Press, New York, pp. 657–64.

Prince, E. & Treuting, R. G. (1956). The structure of tetragonal copper ferrite. *Acta Crystallographica*, **9**, 1025–8.

Quintanar, C., Fuentes, V., Jiménez, M., Aburto, S. & Valenzuela, R. (1986).

Site occupancy of Ga^{3+} in $NiGa_xFe_{2-x}O_4$ ferrites. *Journal of Magnetism and Magnetic Materials*, **54–57**, 1339–40.

Rado, G. T. & Ferrari, J. M. (1975). Electric field dependence of the magnetic anisotropy energy in magnetite (Fe_3O_4). *Physical Review B*, **12**, 5166–73.

Rieck, G. D. & Driessens, F. C. M. (1966). The structure of manganese–iron–oxygen spinels. *Acta Crystallographica*, **20**, 521–5.

Robertson, J. M. & Pointon, A. J. (1966). The cation distribution in nickel ferrite. *Solid State Communications*, **4**, 257–9.

Romeijn, F. C. (1953). Physical and crystallographical properties of some spinels. *Philips Research Reports*, **8**, 304–42.

Röschmann, P. & Hansen, P. (1981). Molecular field coefficients and cation distribution of substituted yttrium iron garnets. *Journal of Applied Physics*, **52**, 6257–69.

Samaras, D., Collomb, A., Hadjivasiliou, S., Achilleos, C., Tsoukalas, J., Pannetier, J. & Rodriguez, J. (1989). The rotation of magnetization in the $BaCo_2Fe_{16}O_{27}$ W-type hexagonal ferrites. *Journal of Magnetism and Magnetic Materials*, **79**, 193–201.

Shannon, R. D. (1976). Revised effective ionic radii and systematic studies of interatomic distances in halides and chalcogenides. *Acta Crystallographica* **A32**, 751–4.

Shepherd, J. P., Aragón, R., Koenitzer, J. W. & Honig, J. M. (1986). Changes in the nature of the Verwey transition in nonstoichiometric magnetite (Fe_3O_4). *Physical Review B*, **32**, 118–19.

Shirane, G., Chikazumi, S., Akimitsu, J., Chiba, K., Matsui, M. & Fuji, Y. (1975). Neutron scattering from low temperature phase of magnetite. *Journal of the Physical Society of Japan*, **39**, 949–57.

Shull, C. G., Wollan, E. O. & Koehler, W. C. (1951). Neutron scattering and polarization by ferromagnetic materials. *Physical Review*, **84**, 912–21.

Smit, J. & Wijn, H. P. J. (1961). *Les Ferrites*. Bibliothèque Technique Philips, Paris.

Suchow, L. & Kokta, M. (1972). Magnetic properties of neodymium-small rare earth ions in two crystallographic sites. *Journal of Solid State Chemistry*, **5**, 85–92.

Sugimoto, M. (1989). Research trends of magnetic amorphous oxides. In *Advances in Ceramics: Proceedings of the Fifth International Conference on Ferrites, India, 1989*, Vol. 1. Eds. C. M. Srivastava and M. J. Patni. Oxford & IBH Publishing Co PVT Ltd, Bombay, pp. 3–12.

Sujata, K. & Mason, T. O. (1992). Kinetics of cation redistribution in ferrospinels. *Journal of the American Ceramic Society*, **75**, 557–62.

Thomson, G. K. & Evans, B. J. (1990). Magnetic properties, composition and microstructure of high-energy product strontium hexaferrites. *Journal of Applied Physics*, **67**, 4601–3.

Thravendrarajah, A., Pardavi-Horvat, M. & Wigen, P. E. (1990). Photoinduced absorption in calcium-doped yttrium iron garnet. *Journal of Applied Physics*, **67**, 4941–3.

References

Turilli, G., Licci, F. & Rinaldi, S. (1986). Mn^{2+}, Ti^{4+} substituted barium ferrite. *Journal of Magnetism and Magnetic Materials*, **59**, 127–31.

Umemura, S. & Iida, S. (1979). Mössbauer study of Fe^{2+} in the low temperature phase of Fe_3O_4. *Journal of the Physical Society of Japan*, **47**, 458–67.

Verwey, E. J. W. (1935). The crystal structure of γ-Fe_2O_3 and γ-Al_2O_3. *Zeitung Kristallographie*, **91 A**, 65–9.

Verwey, E. J. W. (1939). Electronic conduction of magnetite (Fe_3O_4) and its transition point at low temperature. *Nature (London)*, **144**, 327–8.

Verwey, E. J. W., de Boer, F. & van Santen, J. H. (1948). Cation arrangement in spinels. *Journal of Chemical Physics*, **16**, 1091–2.

Verwey, E. J. W. & Heilmann, E. L. (1947). Physical properties and cation arrangements of oxides with spinel structure. *Journal of Chemical Physics*, **15**, 174–80.

Watanabe, A., Yamamura, H., Moriyoshi, Y. & Shirasaki, S. (1982). Crystal chemistry of the spinel-type ferrite series $Li_2M^{4+}Fe_6O_{12}$ ($M^{4+} = Ti^{4+}$, Sn^{4+}, Ge^{4+}, Si^{4+}). In *Ferrites: Proceedings of the Third International Conference*. Eds. H. Watanabe, S. Iida and M. Sugimoto. Center for Academic Publications, Tokyo, pp. 170–3.

Wiesendanger, R., Shvets, I. V., Bürgler, D., Tarrach, G., Güntherodt, H. J., Coey, J. M. D. & Gräeser, S. (1992). Topographic and magnetic-sensitive scanning tunneling microscope study of magnetite. *Science*, **255**, 583–6.

Winkler, G. (1971). Crystallography, chemistry and technology of ferrites. In *Magnetic Properties of Materials*. Ed. J. Smit. McGraw-Hill, London, pp. 20–63.

Wojtowicz, P. J. (1959). Theoretical model for tetragonal-to-cubic phase transformations in transition metal spinels. *Physical Review*, **116**, 32–45.

Wu, C. C. & Mason, T. O. (1981). Thermopower measurements of cation distribution in magnetite. *Journal of the American Ceramic Society*, **64**, 520–2.

Yamada, T., Suzuki, K. & Chikazumi, S. (1968). Electron microscopy of orthorhombic phase in magnetite. *Applied Physics Letters*, **13**, 172–4.

Yao, T., Imafuji, O. & Jinno, H. (1991). EXAFS study of cation distribution in nickel aluminate ferrites. *Journal of the American Ceramic Society*, **74**, 314–17.

3 Preparation of magnetic ceramics

3.1 Introduction

Ferrites can be prepared by almost all the existing techniques of solid state chemistry, leading to a very wide variety of forms: polycrystalline aggregates, thin and thick films, single crystals. Some of these methods have been developed to prepare ferrites with specific microstructures. The oldest one, the *ceramic* method, involves the same operations as the classical techniques for fabrication of conventional ceramics. This explains the origin of the term *magnetic ceramics*. Many of the techniques recently developed consist of improvements in one or several of the basic operations of ceramic fabrication.

The four basic operations in the ceramic method are shown schematically in Fig. 3.1. Raw materials are usually iron oxide (α-Fe_2O_3) and the oxide or the carbonate of the other cations in the desired ferrite. These combine according to the overall reactions:

$$Fe_2O_3 + MeCO_3 \rightarrow MeFe_2O_4 + CO_2 \uparrow \qquad (3.1)$$

$$5Fe_2O_3 + 3R_2O_3 \rightarrow 2R_3Fe_5O_{12} \qquad (3.2)$$

$$6Fe_2O_3 + BaCO_3 \rightarrow BaFe_{12}O_{19} + CO_2 \uparrow \qquad (3.3)$$

for spinels, garnets and hexagonal ferrites, respectively. Me represents a divalent cation and R a rare-earth, trivalent cation. The Fe_2O_3 and the other carbonates or oxides are generally mixed and milled in ballmills; this operation can be carried out in liquid suspension (water, alcohol) to promote a better mixing of the raw materials. After drying, the powder is compacted by dry-pressing in steel dies to obtain the greenbody in the desired, final form, but with slightly greater dimensions, because during the next operation, sintering, shrinkage takes place. To facilitate compaction and to increase the strength of the greenbody, a binder is usually added before pressing.

Sintering is carried out in the solid state, at temperatures between 1000 and 1400 °C, for times of typically 2–24 hours, and in atmospheres ranging from pure O_2 to pure N_2, depending on the ferrite composition and sintering temperature. Reaction (3.1), (3.2) or (3.3), depending on the particular composition, takes place during sintering as well as the process of elimination of voids, or *porosity*, between the particles in the greenbody, leading to an increased density. The initial particle size is critical since the *driving force* for sintering is the reduction in *surface free energy* of the powder. A finishing process is usually needed to obtain a final piece with the exact required dimensions, and is carried out by mechanical machining with diamond tools.

Since ferrites possess a very high melting temperature, both reaction and densification usually take place in the solid state. Initial formation of the product occurs at the contact surface between particles of the two reactants. As reaction proceeds, the product layer becomes thicker, increasing the length of the diffusion paths of the reactants and thus

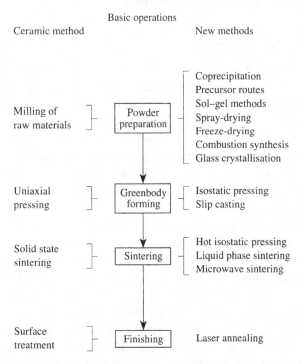

Fig. 3.1. The four basic operations in ferrite preparation. Some components of the ceramic method are on the left; on the right, some of the new methods are indicated.

decreasing the reaction rate. The main diffusion mechanism at this stage is volume (or bulk) diffusion, which depends on the point-defect structure of the solid. A simple method of increasing reaction rates is to crush, mill and repress the product of a first thermal treatment (the *prefiring* stage), and perform a second sintering. By these means, fresh contact surfaces between the reactants are created; this technique also leads to more homogeneous products. In many cases, however, this method is not the best solution, as repeated and long-term milling may introduce defects and impurities into the powder, thereby affecting the properties of the final product.

Many new methods have been developed to overcome the limitations of the ceramic technique. Most involve an improvement in one of the four basic operations in the ceramic method; they appear on the right-hand side in Fig. 3.1. A common feature of the new techniques for powder preparation is that, instead of providing a very small initial particle size by a *physical* change (milling) of the raw materials, the approach is to use *chemical* methods to produce fine, synthetic powders.

3.2   Powder preparation

In the ceramic method, the powder is usually prepared from raw mineral oxides or carbonates by crushing, grinding and milling. The most common type of mill is the ballmill, which consists of a lined pot with hard spheres or rods inside. Milling occurs by impact of the spheres or rods on the powder. There is a limiting rotation speed for an efficient tumbling, imposed by centrifugal forces. Milling can be carried out in a wet medium to increase the degree of mixing, by forming an aqueous or alcoholic suspension of the raw materials. After milling, the suspension is dried to obtain the powder.

Mechanical milling cannot produce fine powders; even after long-term milling, the limiting particle size is $\sim 0.2 \,\mu m$. Extended milling has several shortcomings: the spheres of the mill become worn and introduce significant quantities of undesired impurities into the powder; the distribution in particle size becomes extremely wide. Such characteristics are different from those expected for an ideal powder, Table 3.1.

A small particle size of the reactant powders provides a high contact surface area for initiation of the solid state reaction; diffusion paths are shorter, leading to more efficient completion of the reaction. Porosity is easily eliminated if the initial pores are very small. A narrow size

Table 3.1. *Ideal characteristics of ferrite powders.*

(1) Small particle size (submicron)
(2) Narrow distribution in particle size
(3) Dispersed particles
(4) Equiaxed shape of particles
(5) High purity
(6) Homogeneous composition

distribution of spherical particles as well as a dispersed state are important for compaction of the powder during greenbody formation. Grain growth during sintering can be better controlled if the initial size is small and uniform.

Synthetic powders for most ferrite compositions, with practically all the ideal characteristics in Table 3.1, can be prepared by various of the new methods. Their common feature is that the mixing of the components takes place at the *atomic*, or *molecular* scale. In one group of these techniques, a solution is formed with the cation stoichiometry of the desired ferrite; a controlled precipitation (coprecipitation) leads to a mixture of very small particles containing the cations in the right proportion, in the form of hydroxides, for instance. The particle size can be accurately controlled by the pH of the solution. In another group, precipitation occurs by removal of the solvent by either evaporation (spray-drying) or solidification (freeze-drying), or by the addition of a different solvent (sol–gel). In the case of a molecular precursor, the technique is based on the preparation of a chemical compound, an organic molecule containing the cations of the ferrite in the desired proportions (organic precursor). In all cases, the ferrite particles are obtained by heating the precipitate.

A common problem in all these methods is the tendency to agglomeration of particles during precipitation or removal of the solvent, since some suspensions are in the colloidal range (particles with at least one dimension smaller than ~ 200 nm). *Agglomerates* are associations of individual particles and are undesirable in the initial powder because they lead to inhomogeneous greenbodies and defects in sintered pieces. A useful model of the agglomeration forces is provided by the DVLO theory (for Deryaguin, Verwey, Landau and Overbeck) where attractive forces are of the van der Waals type and the repulsion occurs by the presence of an electrostatic double-layer surrounding each particle (McColm & Clark, 1988).

3.2.1 Coprecipitation

This method of ferrite formation is based on the formation of aqueous solutions of chlorides, nitrates or sulphates of Fe^{3+}, and of divalent Ni, Co, Mg, Ba, Sr, etc., in the concentrations required for the ferrite composition, and their *simultaneous precipitation* in the form of hydroxides by NaOH. The precipitate is then filtered, washed and dried. The ferrite particles are obtained by calcination at 180–300 °C in air. Particles with a narrow size distribution in the range 50–500 nm may be obtained, with high purity. The final ferrite body is obtained by sintering at temperatures considerably lower than in the case of the ceramic method.

A common source of inhomogeneity in the ferrite product is associated with changes of concentration and pH during precipitation. The various metal hydroxides can have different solubility dependences on pH, and the cation ratios in the precipitate can be different from those in the solution. To avoid this problem, the solution of metal ions is usually added to the alkaline solution slowly and dropwise (McColm & Clark, 1988). If instead the alkaline solution is added to the solution of metal ions, the pH of the suspension must be brought to a value above 10 to avoid differences in precipitate composition (Takada & Kiyama, 1971).

Ferrites can also be synthesised by the coprecipitation of oxalates. The hydrated iron oxalate, $FeC_2O_4 \cdot 2H_2O$ and the oxalates of Mn, Co, Ni or Zn can be coprecipitated from solution by solvent evaporation (Wickham, 1967). The Fe/Me^{2+} stoichiometry can be accurately controlled to within 1%.

Iron hydroxide, $Fe(OH)_3 \cdot nH_2O$, decomposes with temperature to γ-Fe_2O_3, which is more reactive than α-Fe_2O_3. As a result, strontium hexaferrite, $SrFe_{12}O_{19}$, can be obtained from coprecipitates of $Fe(OH)_3 \cdot nH_2O$ and strontium laureate, $Sr[CH_3(CH_2)_{10}COO]_2$, at temperatures as low as 550 °C (Qian & Evans, 1981). By contrast, mechanical mixtures of α-Fe_2O_3 and SrO show an appreciable reaction rate only at $T > 720$ °C.

Some ferrites that have been obtained by coprecipitation are shown in Table 3.2. Besides the preparation of powders for the ferrite industry, raw materials prepared by coprecipitation are used for the manufacture of pigments and magnetic toners and for the removal of heavy metal ions from waste water (Takada, 1982).

Table 3.2. *Some ferrites obtained by coprecipitation methods.*

Ferrite	Salts	Precipitant	References
Spinels Mn, Zn, Co, Mg, Fe^{2+}	Sulphates	NaOH	Takada & Kiyama (1971)
Spinels Mn, Zn, Co, Mg, Fe^{2+}	Oxalates	Solvent evaporation	Wickham (1967)
Spinels Mn–Zn	Sulphates	$NaOH + NH_4HCO_3$	Yu & Goldman (1982)
Ba–M	Chlorides	NaOH	Hibst (1982)
Sr–M	Chlorides	NaOH	Date *et al.* (1989)
Sr–M	Nitrates	Lauric acid	Qian & Evans (1981)

3.2.2 Precursor methods

The precursor method allows the preparation of ferrites with a precise stoichiometry. It involves the synthesis of a compound, the *precursor*, in which the reactants are present in the required stoichiometry; upon heating, the precursor decomposes yielding the ferrite.

Spinel ferrites have been prepared by thermal decomposition of crystallised pyridinates $Me_3Fe_6(CH_3COO)_{17}O_3OH \cdot 12C_5H_5N$, with Me = Mg, Mn, Co and Ni (Wickham, Whipple & Larson, 1960). Pyridinate crystals were prepared by dissolution of the basic double acetate, $Ni_4Fe_9(CH_3CO_2)_{26}(OH)_9 \cdot 23H_2O$, in hot pyridine, and then precipitated by solvent evaporation. To increase the purity of the salts, a double recrystallisation process was performed. To obtain the ferrite, the products were ignited at high temperature in air to yield the correct O content. For most of the ferrites, this temperature was 1000 °C – an exception was Mn ferrite, which required an ignition temperature of 1300 °C.

Ni–Zn ferrites have been prepared by a hydrazinium metal hydrazine-carboxylate precursor, $(N_2H_5)_3Ni_{1-x}Zn_xFe_2(N_2H_3COO)_9 \cdot 3H_2O$, with $0 \leq x \leq 1$ (Srinivasan *et al.*, 1988). Preparation of the precursor involved reaction of aqueous solutions of the Fe, Ni and Zn sulphates with a solution of N_2H_3COOH in $N_2H_4 \cdot H_2O$. The crystalline precursor was obtained in ~ 2 days, filtered, washed, dried and calcined in O_2 at 250 °C. The size of the ferrite particles was 20–60 nm, with a high purity.

Using a similar technique, Chen *et al.* (1988) have produced ultrafine particles of Ni–Zn ferrites in the 3–30 nm range. A solution of the nitrates of Ni, Fe and Zn was precipitated by addition of N_2H_4 and KOH. The gelatinous precipitate was aged at 90 °C for 1 h, filtered, washed and dried in vacuum at 50 °C. Particles with high magnetisation were obtained by annealing at 150 °C.

3.2.3   Sol–gel methods

Sol–gel techniques are receiving much attention because they can be applied to an extremely wide variety of materials; they offer the possibility of controlling not only the size and distribution of particles, but also their shape. Fibres, films and monoliths can also be obtained by sol–gel methods.

The process involves the preparation of a *sol*, which is a dispersion of solid particles, the *dispersed phase*, in a liquid, the *dispersed medium*; at least one dimension of the particles of the dispersed phase is between 1 nm and 1 µm. The sol is prepared by mixing concentrated solutions containing the cations of interest, with an organic solvent as the dispersion medium. It is also possible to begin with a colloidal suspension, instead of, or in addition to, the solutions. The sol is then destabilised, generally by adding water. The presence of water modifies the pH of the sol and reduces the repulsion between particles. This results in a large increase in the viscosity of the system, leading to formation of a *gel*.

Depending on the amount and rate of water uptake, two different gels may be obtained as a result of the destabilisation of the sol, Fig. 3.2. If excess water is used, the gel is a continuous network of aggregates formed by the particles; it is a *colloidal gel*, or an *aquagel*. If the water is added slowly and in small amounts, the sol particles increase in size by a condensation–polymerisation reaction. This gives a *polymeric gel*, also known as an *alcogel*. The next stage is the removal of the dispersion medium. Again, the rate at which the operation is performed is important. Fast removal leads to a powder; this can be performed by passing fine droplets of the gel through a column of an alcohol. The final ceramic can be obtained by compaction and sintering of the *powder xerogel* at an appropriate temperature. On the other hand, a very slow and controlled removal of the liquid can lead to fabrication of fibres, coatings and monoliths.

The size of the particles in the powder xerogel depends on the type of organic solvent, the temperature and the rate of removal of the liquid. The process can be operated on a continuous basis, Fig. 3.3. Its only present disadvantage is the high cost of the materials used.

Powder preparation

Ba hexaferrite has been prepared from Fe and Ba propoxides, $Fe(OC_3H_7)_3$ and $Ba(OC_3H_7)_2$, respectively (McColm & Clark, 1988). The solvent used to form the sol was C_3H_7OH. After the addition of water, the alcogel was aged for 12–24 h; the particles were separated by centrifugation and dried at 100 °C to obtain an amorphous powder. The Ba hexaferrite was obtained by calcining at 700 °C for 2 h, with a final particle size of about 1 μm.

A variety of ferrites have been prepared from colloidal suspensions (Fan & Matijevic, 1988). Sr hexaferrite was crystallised from a $Fe(OH)_2$ gel in the presence of $SrCl_2$, with KNO_3 as a mild oxidising agent at 90 °C. The size of the particles was 0.08–0.51 μm.

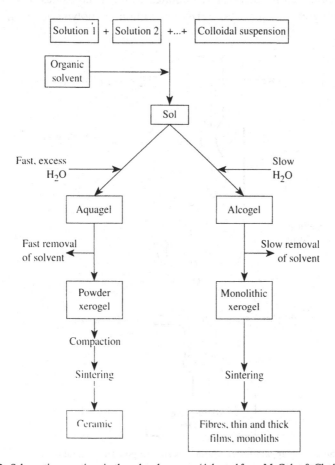

Fig. 3.2. Schematic operations in the sol–gel process. (Adapted from McColm & Clark, 1988.)

Substituted YIGs have been prepared by sol–gel methods (Matsumoto, Yamaguchi & Fujii, 1991), with a solution of nitrates as the initial sol. The gel reaction was promoted by adding a small amount of ethylene glycol. After evaporation of the alcohol at 80 °C, the dry gels were heated at 110 °C in air to eliminate NO_x and calcined at various temperatures up to 900 °C. Formation of YIG and Bi-substituted yttrium garnet, $Bi_xY_{1-x}Fe_5O_{12}$ (with $x \leq 1.7$), was noticeable for $T > 700$ °C; for the latter, synthesis temperature tended to decrease with Bi content.

Sol–gel methods have been used to produce a coating of additives on ferrite particles prior to sintering, in an attempt to achieve a good dispersion (Brooks & Amarakoon, 1991). The coating was a borosilicate obtained from metal alcoholates or metal alkoxides. Hydrolysis was catalysed by an acid (HF) to obtain a polymeric chain and

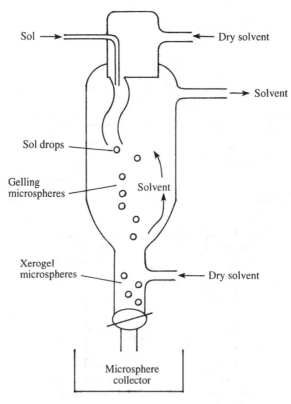

Fig. 3.3. Schematic apparatus for continuous sol–gel powder preparation. (Adapted from McColm & Clark, 1988.)

Powder preparation

facilitate coating formation. The coating thickness was in the range 5–15 nm.

3.2.4 Spray-drying

Precipitation from a concentrated solution of cations can be performed by solvent evaporation. To ensure that the particle size remains small, the concentrated solution may be atomised at high pressure into fine droplets of 100–500 μm diameter; the solvent is rapidly evaporated by an upward stream of hot gas. The particles obtained, which can be as small as 100 nm, are compacted and calcined to produce the ceramic. A schematic representation of the spray-drying process is shown in Fig. 3.4. Several alternative methods are currently under development; they are known as *aerosol synthesis*, *aerosol pyrolysis* or *mist pyrolysis*, depending on the specific technique to produce the gaseous suspension of fine particles; aerosols are produced in high pressure nozzles and mists are obtained by means of nebulisers. YIG particles (0.25 μm) have been obtained by mist pyrolysis (Matsumoto *et al.*, 1991) by nebulising an aqueous solution of

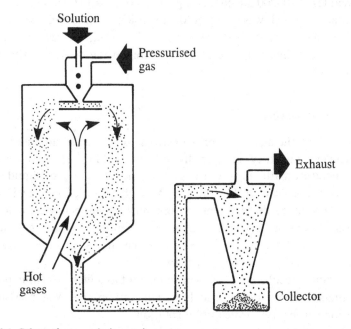

Fig. 3.4. Schematic spray-drying equipment.

the corresponding nitrates; the generated mist was fed to a reaction tube at 800 °C, under N_2 gas as carrier. The reaction time was about 20 s. The collected particles were amorphous; crystalline particles were obtained by calcination at $T \geq 950$ °C. Aerosol methods have been used to prepare Ba hexaferrite (Kaczmarek, Ninham & Calka, 1991) and gadolinium iron garnet, $Gd_3Fe_5O_{12}$ (Xu *et al.*, 1992); by varying the reaction conditions, particle sizes 0.05–1 µm were obtained in both cases.

Mn–Zn ferrites have been prepared by calcination at 900 °C of spray-dried particles (Jain, Das & Avtar, 1976). The solution was prepared by mixing sulphates. The formation of the Mn–Zn ferrite involved decomposition of the $FeSO_4$ to α-Fe_2O_3, which reacted with the $ZnSO_4$ to form Zn ferrite. $MnSO_4$ decomposed last to react with the Zn ferrite leading to the desired Mn–Zn ferrite.

An alternative method that has been used to prepare ferrite particles in a single step is the evaporative decomposition of the solutions, or *spray-roasting* (Ruthner, 1977). The solutions were mixed and atomised, and the droplets fell through a reaction chamber at 900–1050 °C. The solvent evaporated and the salts decomposed to oxides. The process took 3–5 s. By using a roasting furnace for industrial production, agglomerates of 40–200 µm containing 1 µm ferrite particles were obtained. The furnace feedstock was an aqueous suspension of the oxides, carbonates or hydroxides of the desired composition. However, the residence time was insufficient and complete transformation to the desired ferrite was not achieved.

3.2.5 Freeze-drying

In this method, the aqueous, concentrated solution is also atomised into fine droplets, but they are rapidly frozen by blowing them into a low-temperature bath, such as ice–acetone, liquid C_6H_{14} or liquid N_2. The droplets are then dried in vacuum, by sublimation of the ice without melting. This means that the temperature must remain below the eutectic in the salt–H_2O system. The anhydrous salts (nitrates, sulphates, chlorides, etc.) are calcined to produce powders which are ~ 0.1 µm in size. A schematic freeze-drying process is shown in Fig. 3.5.

A non-aqueous solvent can also be used, if it has a higher melting point and a higher vapour pressure at low temperatures than water. Also, a suspension can be used instead of a solution.

Ni–Zn ferrites have been obtained by freeze-drying; a high density with

Powder preparation

small, uniform grain size was obtained in the final ferrite body when the freeze-dried powders were compacted by hot isostatic pressing (Schnettler & Johnson, 1971).

3.2.6   Combustion synthesis

A novel method for the preparation of fine particles of ferrites makes use of the strong *exothermic redox reaction* between metal nitrate (oxidants) and tetraformal trisazine, $C_4H_{16}N_6O_2$ (TFTA), or oxalic acid dihydrazine, $C_2H_6N_4O_2$ (ODH), fuels (Suresh, Kumar & Patil, 1991). In this process, the stoichiometric ratio of nitrates is dissolved in the minimum amount of water in a pyrex dish; the fuel (TFTA or ODH) is added and the dish is heated at 350 °C in a muffle furnace. After boiling and ignition of the mixture, a spinel residue is obtained in a few minutes. A heating rate of at least 75 °C/min is used to obtain good combustion. This method has been used for synthesis of Ni–Zn and Co spinels; in the case of Cu ferrite, the ignition of the fuel–oxidant took place at room temperature. Single phase YIG was obtained by heating the combustion residue at 850 °C for

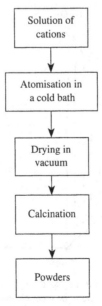

Fig. 3.5. Steps in the freeze-drying process.

6 h. The observed particle size was 10–30 nm; the ODH fuel led to smaller particles than the TFTA process.

3.2.7   Glass recrystallisation

This technique for preparing small particles of M- and W-type hexaferrites involves the synthesis of an intermediate *glassy* phase in the pseudo-ternary phase diagram Fe_2O_3–BaO–B_2O_3, and the crystallisation of the ferrite particles from the amorphous phase. Crystallisation takes place in the Fe_2O_3–BaB_2O_4–$Ba_2Fe_2O_5$ triangle, Fig. 3.6 (Haberey, 1987). The lowest temperature of primary crystallisation for barium hexaferrite is close to 800 °C for the composition 55 mole% BaO, 20 mole% B_2O_3 and 25 mole% Fe_2O_3. The glass is prepared by melting the components in air at temperatures between 1000 and 1300 °C. To obtain a glassy phase, the melts are quenched between two Cu blocks (splat-cooling). Ferrite particles are obtained as small crystals in the glass matrix by heat treatment and separated by leaching and washing.

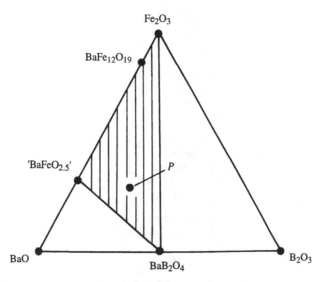

Fig. 3.6. Pseudo-ternary phase diagram Fe_2O_3–BaO–B_2O_3. Hexaferrites are obtained in the triangle Fe_2O_3–BaB_2O_4–'$Ba_2Fe_2O_5$'. The composition P has the lowest temperature of primary crystallisation. (Adapted from Haberey, 1987.)

3.2.8   Other methods

There are two other methods for producing ferrite powders that deserve attention: molten-salt synthesis and shock-wave loading. In *molten-salt* synthesis, Fe_2O_3 and the corresponding carbonate, oxide, hydroxide or nitrate needed to produce the ferrite phase are dry blended with a mixture of NaCl and KCl. This reaction mixture, in a Pt crucible, is placed in a furnace at 800–1100 °C for ~1 h; the chloride solvent is melted and provides an efficient heating medium for the reaction. After cooling, the solvent is separated from the ferrite by dissolution in water. The product can be finally collected by filtration. Ba and Sr hexaferrite powders have been prepared by this method (Arendt, 1973). Submicron, high-quality crystallites with a low ferrous content were obtained.

Shock synthesis consists of subjecting the material to the high-pressure shock-wave loading from an explosion. The increase in solid-state reactivity is attributed to the introduction of large concentrations of defects into the material. Zn ferrite has been prepared by this method (Venturini, Morosin & Graham, 1985). Fe_2O_3 and ZnO were used as reagents; after ballmilling, the stoichiometric mixture was pressed and placed in sealed Cu capsules to various initial packing densities, and subjected to controlled explosive loading. The maximum ferrite fraction formed was 27 wt%. The packing density resulted in a decrease in peak shock temperature: densities of 2.15×10^3 and 3.30×10^3 kg/m^3 resulted in shock peak temperatures of 655 and 300 °C, respectively. The Zn/Fe ratio in the spinel phase was different from that in the initial mixture of oxides.

3.3   Greenbody forming

3.3.1   Compaction

In this operation, the powder is compacted into a piece, the *greenbody*, usually having the same shape as the final piece. The aim of greenbody forming is to achieve a high-density particle packing. Two particular aspects of the degree of compaction are critical for the subsequent operation (sintering): a high contact surface area and a small porosity. Diffusion depends on the contact surface area; small pores are easier to eliminate during sintering. Greenbody dimensions are always larger than those of the final piece because during the high-temperature sintering, *shrinkage* occurs.

Particle packing density of a greenbody is related to its mechanical strength, or *greenbody strength*. By assuming that the particles are spheres, and the attractive forces are of the van der Waals type, an approximate relation has been derived to estimate greenbody strength (McColm & Clark, 1988):

$$\sigma_g = \frac{A'\phi_v}{(1 - \phi_v)rs_p^2} \tag{3.4}$$

where σ_g is the greenbody strength, ϕ_v the solid volume fraction, A' a constant, and r and s_p are the radius of and the separation between particles, respectively. Relation (3.4) is also known as the Rumpf equation. The greenbody strength thus strongly depends on particle packing through ϕ_v and s_p, and is also favoured by a small particle size, r. A wide distribution in particle size allows a better packing; however, due to its effects on grain growth during sintering, a narrow distribution of small particles is preferable. Greenbody strength increases with the pressure applied during compaction by decreasing s_p; however, excessive pressure leads to differences in density which result in macroscopic defects. Experimental strength values are always lower than the values predicted by Eq. (3.4). A more realistic approach is based on the existence of defects in the greenbody, such as cracks and packing inhomogeneities (Kendall, 1984).

As pressure increases, compaction of a dry powder consisting of agglomerates made of small particles occurs in the following steps:

(1) Agglomerates are brought together; spaces between them are filled.
(2) Fragmentation of the agglomerates occurs.
(3) Spaces between particles are filled.
(4) The particles are fragmented.

In an ideally dispersed powder consisting only of isolated particles, the sequence reduces to steps (3) and (4).

The sequence of compaction steps has been observed in ferrite powders (Youshaw & Halloran, 1982); the early stage of compaction additionally included plastic deformation of the agglomerates. Their ductility plays an important role in the process. The pressing rate was observed to be important; compaction was retarded at faster pressing rates.

A *binder* is usually added prior to compaction, at a concentration lower than 5 wt%. Binders are polymers or waxes; the most commonly used in ferrites is polyvinyl alcohol. The binder facilitates the particle flow during compacting and increases the bonding between particles, presumably by forming bonds of the type particle–binder–particle. During sintering, binders decompose and are eliminated from the ferrite. In the case of Mn–Zn ferrites, burning of the binding can have a significant effect on the cation oxidation states at the surface of the ferrite (Broussaud *et al.*, 1989). Reduction occurs which can lead to improved greenbody strength. Binders can also have long-term effects; degradation of the magnetic properties of Mn–Zn ferrites has been attributed to the interaction of the binder used in ferrite processing with common airborne industrial contaminants (Ghate, Holmes & Pass, 1982).

3.3.2 Pressing

Uniaxial pressing is the simplest compaction technique; it consists of the compression of the dry powder between a die and punch system, by applying uniaxial pressure usually ~ 5–150 MPa. An almost unavoidable problem is that friction between the powder and the die walls produces an inhomogeneous flow of particles, Fig. 3.7, resulting in pressure and density gradients in the greenbody. Sections *A* in Fig. 3.7 experience the highest friction and are thus subjected to the highest pressure. The shear stress appears along faces a_1 and a_2, as shown in Fig. 3.8. An increase in applied pressure usually results in a higher difference in local densities. These differences in compaction can be high enough to produce *capping*, which is the spontaneous separation of top layers of the greenbody after pressing, and sometimes after sintering. Fine powders (<0.1 µm) are sometimes difficult to press, because they agglomerate easily and tend to form inhomogeneous packing arrangements (McColm & Clark, 1988).

Isostatic pressing has been developed to avoid the frictional problems associated with uniaxial pressing. The powder is placed in a flexible mould and compaction takes place by immersing in a fluid under pressures as high as 100–250 MPa. Since compaction takes place homogeneously, a uniform density can be obtained in complex shapes and with smaller amounts of binder, as compared with uniaxial pressing. The process typically includes several pressing stages, starting with lower pressures. Powder requirements for this technique are free-flow and non-agglomeration, which can be met by means of the new methods such as spray-drying (McColm & Clark, 1988). An important variation of this pressing technique is *hot isostatic pressing* (HIP), where pressure and heat are applied simultaneously. HIP is described in Section 3.4.3.

3.3.3   Casting

Casting is a technique used in traditional ceramics; it consists of pouring an aqueous, colloidal suspension, called a *slip*, into a mould which slowly absorbs the liquid by filtration through its walls, producing a uniform sedimentation of the particles. After a certain time, the excess slip is removed, the piece is separated from the mould, dried and fired. The mould is usually made of gypsum ($CaSO_4$), because it can absorb considerable amounts of water and the moulded piece can easily be separated from it. In addition to the normal requirements for a colloidal suspension, such as stability and homogeneity, the slip should have a low viscosity and produce a piece with small shrinkage on drying. This method has been used to prepare Ni–Zn ferrites (Tseng & Lin, 1989); the slip was

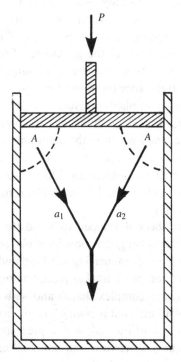

Fig. 3.7. Stress in uniaxial pressing. The friction between particles and the die walls, sections A, results in a high shear-stress along faces a_1 and a_2 (McColm & Clark, 1988).

prepared from chemically coprecipitated Ni–Zn ferrite powders of size ~30 nm. After moulding, the pieces were dried at 50 °C and sintered at 1100 °C, leading to a density 98% of the theoretical value.

A related technique used to produce very thin, virtually continuous tapes (0.1 mm) of ceramic materials is *tape casting*. This involves the preparation of a colloidal suspension which is transported on a carrier

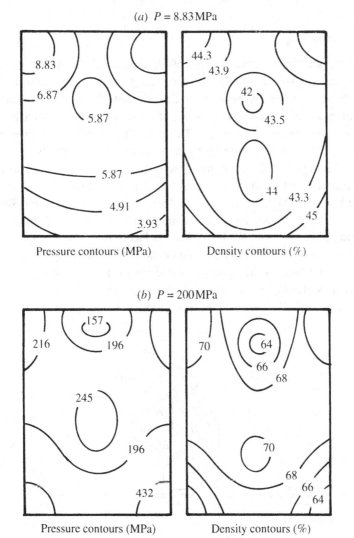

(a) $P = 8.83$ MPa

Pressure contours (MPa)   Density contours (%)

(b) $P = 200$ MPa

Pressure contours (MPa)   Density contours (%)

Fig. 3.8. Pressure and density contours in a uniaxially pressed powder at a maximum pressure of: (*a*) 8.83 MPa, (*b*) 200 MPa (McColm & Clark, 1988).

61

belt, Fig. 3.9. The thickness of the suspension is controlled by a double system of doctor blades to obtain a precise layer thickness. The belt is made of a permeable material to absorb the liquid of the suspension and initiate the drying process. Once dried, the tapes are sintered. Tapes with good surfaces and uniform thickness can be obtained by this method.

3.4   *Sintering*

Sintering is defined as the process of obtaining a dense, tough body by heating a compacted powder for a certain time at a temperature high enough to significantly promote diffusion, but clearly lower than the melting point of the main component. It is an extremely complex process with many interacting variables, Table 3.3. The ideal sintering process results in a fully dense material by elimination of the porosity.

The driving force for sintering is *the reduction in surface free energy* of the powder. Part of this energy is transformed into interfacial energy (grain boundaries) in the resulting polycrystalline body (Kingery, Bowen & Uhlman, 1976). An important contribution to the understanding of sintering phenomena in the last few years has been the recognition of the role of surfaces and interfaces, which determine not only the macroscopic driving force, but also the microscopic diffusion mechanisms.

Several sintering mechanisms are possible, depending on the phases present, Fig. 3.10. The possibilities include: *liquid-phase, viscous-flow,*

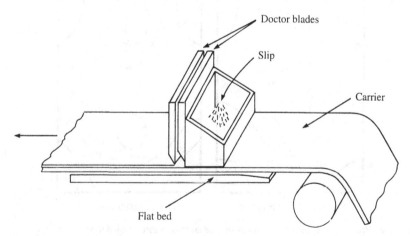

Fig. 3.9. Schematic representation of the tape casting process (McColm & Clark, 1988).

Table 3.3. *Variables in sintering.*

Powder	Composition
	Average particle size
	Particle size distribution
	Particle shape
	Compaction (greenbody density)
Impurities	Melting temperature
	Solubility in product phase
	Cation valencies
Conditions	Heating rate
	Temperature
	Time
	Cooling rate
	Oxygen partial pressure

solid-state and *reaction* sintering. In liquid-phase sintering, a small fraction of the material melts at the sintering temperature. Densification is promoted by liquid filling the pores and grain boundaries. In the latter, grain rearrangement is facilitated by the presence of the liquid (Kang, Kim & Yoon, 1991). Ceramic materials such as refractories and alumina insulators are prepared by this method (Brook, 1991).

In viscous-flow sintering, some 20% of the initial solid melts at the sintering temperature and the densification of the material occurs by crystallisation of this phase (or vitrification if it becomes a glass) during cooling. Porcelains are prepared by this method.

In solid-state sintering, the material is in the solid phase during the whole process. *Reaction sintering* refers to the chemical reaction and densification of a mixture in a single thermal treatment. In this process, the reduction in free energy due to the chemical reaction is an additional driving force. It also has economic advantages in terms of time and fuel costs, since fewer stages are involved. Reaction sintering can lead to a preferential orientation of grains (Stählein & Willbrand, 1972) which can have an important effect on properties.

In the simplest model for representing the elimination of porosity in a solid, Fig. 3.11, volume diffusion takes place by means of lattice defects, in particular, vacancies. Pores play the role of vacancy sources, and grain boundaries act as vacancy sinks. The key point in this mechanism is that a curved surface *creates* a pressure difference between its sides. This may be seen by considering the work, $\Delta P \, dv$, that has to be done to create a pore of volume dv in a solid of surface energy γ, Fig. 3.12:

$$\Delta P \, dv = \gamma \, dA \tag{3.5}$$

For a spherical pore of radius r, $dv = 4\pi^2 \, dr$ and $dA = 8\pi r \, dr$; the pressure difference becomes:

$$\Delta P = 2\gamma/r \tag{3.6}$$

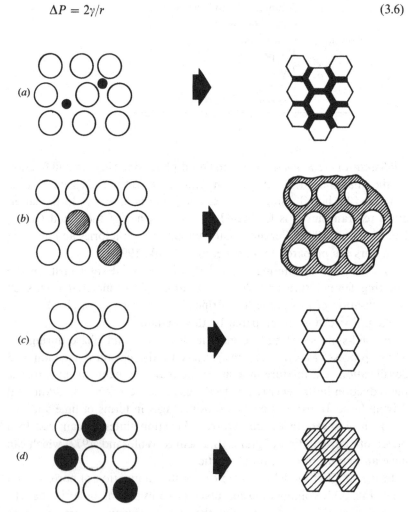

Fig. 3.10. Sintering mechanisms: (*a*) liquid-phase, (*b*) viscous-flow, (*c*) solid-state, and (*d*) reaction sintering. (Adapted from Brook, 1991.)

The pressure difference is thus a function of the radius of curvature. This pressure difference is significant only for a small radius of curvature, typically in the micron range. The differences in local pressures of curved surfaces give rise to chemical potential gradients, which, in turn, influence local defect concentrations (Brook, 1989).

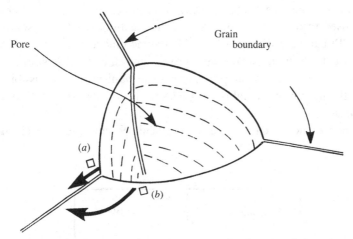

Fig. 3.11. Model for elimination of porosity. Vacancy flux by: (*a*) grain boundary diffusion, (*b*) volume diffusion.

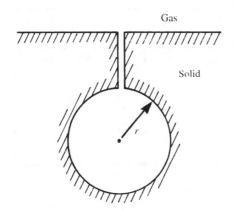

Fig. 3.12. Hypothetical creation of a pore of radius r and surface energy γ. (Adapted from Kingery *et al.*, 1976.)

3.4.1   Solid-state reactions

An illustrative and now classical example of methods for elucidating material transport mechanisms in solid-state reactions is the use of couple arrangements, Fig. 3.13 (Wagner, 1936; Reijnen, 1991). Couple experiments are helpful for illustrating the complexity of reaction mechanisms in ionic compounds in general, and ferrites in particular.

The diffusion couple consists of two reactant pieces placed in contact. In some cases, the differences in colour of reactants and the product phase can be used to observe the progress of the reaction. In other cases, thin wires of Pt can be used as markers. After some time at high temperature, the product phase nucleates and grows, and therefore separates the reactants. For relatively simple reaction mechanisms, it is possible to deduce the nature of the diffusing species from the relative growth of the new phase.

In practice, these techniques can be difficult. When significant changes

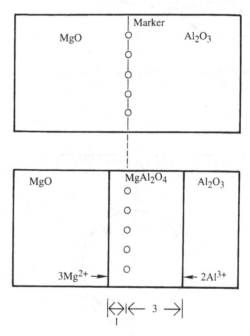

Fig. 3.13. Reaction couple arrangement to study the formation of the spinel $MgAl_2O_4$ from MgO and Al_2O_3.

in density are involved, shrinkage of the product phase may lead to fracture of the reactant–product interface; also, no marker is really inert. Kooy (1965) has proposed several couple geometries to minimise these effects.

A simple example of the use of couple experiments is the formation of the spinel $MgAl_2O_4$ from magnesia, MgO, and alumina, Al_2O_3. The overall reaction is:

$$MgO + Al_2O_3 \rightarrow MgAl_2O_4 \tag{3.7}$$

The relative amounts of the spinel product phase formed at the original boundary are indicative of the diffusion mechanism of the reaction. Experimental results show that the amounts of spinel product are formed in a ratio 1:3 at the marker (Reijnen, 1991), Fig. 3.13.

The reaction may occur by a number of mechanisms, provided electro-neutrality is maintained. An extreme assumption could be, for example, that the only mobile cation is Al^{3+}. Electro-neutrality involves a parallel diffusion of three O^{2-} ions, with every two Al^{3+} ions. An important consequence of such a diffusion mechanism is that the marker would be displaced; both alumina ions would be transported towards the spinel–MgO boundary while the spinel–Al_2O_3 boundary progresses into the alumina side. The displacement of the marker is known as the *Kirkendal effect.*

However, O^{2-} ion diffusion is not very common. Due to their larger ionic radius, O^{2-} ions are expected to have a considerably lower mobility than most cations. When cations are the only transporting species, the mechanism is known as the *Wagner mechanism* (Wagner, 1936). Formation of the spinel phase can be easily explained by diffusion of Mg^{2+} and counter-diffusion of Al^{3+}. To maintain electro-neutrality, the mechanism involves the counter-diffusion of three Mg^{2+} ions for every two Al^{3+} ions (Carter, 1961).

The ratio of spinel formed at the sides of the markers is 1:3, due to the stoichiometry of the spinel phase and the relative amounts of Mg^{2+} and Al^{3+}: while three Mg^{2+} ions produce three $MgAl_2O_4$, the diffusion of two Al^{3+} ions results only in one $MgAl_2O_4$. An additional complication is that, at the high reaction temperatures, the spinel phase exists as a range of solid solutions on the join $MgO-Al_2O_3$, extending from $MgAl_2O_4$ on the MgO-rich side, to $\sim Mg_{0.73}Al_{2.18}O_4$ (depending on temperature) at the Al_2O_3-rich limit. In the initial stages of reaction, the spinel product can hence be inhomogeneous.

The formation of the magnesium ferrite, $MgFe_2O_4$ from MgO and Fe_2O_3 is more complex. Instead of a 1:3 ratio in the product phase at the marker's sides, characteristic of a mechanism involving counter-diffusion of three Mg^{2+} ions and two Fe^{3+} ions, a 1:2.7 ratio has been observed (Carter, 1961). Significantly, the marker was found to be displaced, Fig. 3.14. The reduction of Fe^{2+} to Fe^{2+} and a diffusion mechanism $Fe^{2+} \leftrightarrows Mg^{2+}$ (with reoxidation of Fe^{2+} back to Fe^{3+} in the spinel phase) would lead to a product phase ratio 1:2 at the two interfaces.

The fact that the ratio 1:2.7 is observed indicates that *both Fe^{2+} and Fe^{3+}* are diffusing. The marker displacement is associated with oxygen transport in the gas phase (Kooy, 1965). The loss of oxygen during the initial steps of reaction has also been observed as a loss of weight, which is recovered as reaction is completed, with the associated reoxidation of Fe^{2+} to Fe^{3+}.

The existence of an extensive range of solid solutions can also lead to a difference in rate of consumption of the reactants. Fe_2O_3 has a high

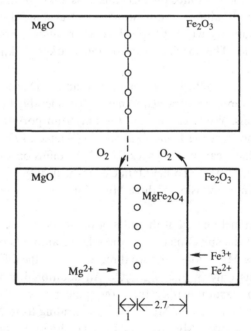

Fig. 3.14. Reaction couple arrangement to study the reaction of MgO and Fe_2O_3 to form $MgFe_2O_4$. Experimentally, a 1:2.7 ratio was observed, as well as a marker displacement (Carter, 1961).

Sintering

solubility in the spinel phase through partial reduction of Fe^{3+} to Fe^{2+} to produce magnetite:

$$3Fe_2O_3 \rightarrow 2Fe^{2+}Fe_2^{3+}O_4 + \tfrac{1}{2}O_2 \uparrow \qquad (3.8)$$

The Fe_2O_3 reactant can therefore be consumed before the MgO has completely reacted. For short reaction times, an inhomogeneous spinel phase may thus result.

Reaction mechanisms become more complex when the reaction to obtain the desired material involves several steps, with the formation of intermediate compounds. An example is the preparation of hexagonal ferrites from Fe_2O_3 and the corresponding carbonates (Winkler, 1965). The compounds $BaFe_3O_4$ (F), $CoFe_2O_4$ (S), $BaFe_{12}O_{19}$ (M), and $Ba_2Co_2Fe_{12}O_{22}$ (Y) were detected as intermediates during the preparation of $Ba_6Co_4Fe_{48}O_{82}$ (Z); at high temperatures Z decomposes and $Ba_2Co_4Fe_{32}O_{54}$ (W) starts to form, Fig. 3.15.

3.4.2   Densification

To obtain a dense, tough polycrystalline aggregate it is necessary to eliminate the voids between the particles of the ferrite and form grain boundaries. Although there is a continuous evolution of the microstructure

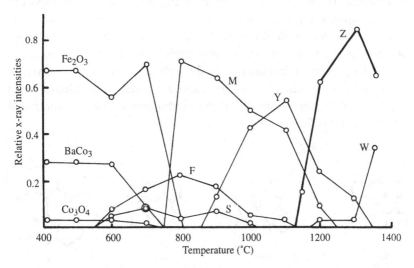

Fig. 3.15. X-ray intensity of the principal diffraction peak of intermediate compounds as a function of temperature, during preparation of $Ba_6Co_4Fe_{48}O_{82}$ (Winkler, 1965).

Table 3.4. *Sintering stages.*

Stage 1	Contact area between particles increases
Stage 2	Porosity changes from open to closed porosity
Stage 3	Pore volume decreases; grains grow

during sintering, it is commonly divided into three stages, Table 3.4 (McColm & Clark, 1988).

In the initial stage, neighbouring particles form a neck by surface diffusion and presumably also at high temperatures by an evaporation–condensation mechanism, Fig. 3.16. Grain boundaries begin to form at the interface between particles with different crystallographic orientation. Pores appear as voids between at least three contacting particles. Shrinkage occurs.

The second stage of sintering begins when a three-dimensional network of necks is achieved. During this stage, most of the densification occurs. To decrease and eventually eliminate pore volume, a net transport of material to the pores by volume diffusion is required. The mobility of the atoms (or ions) is greatly enhanced by the presence of lattice defects. The sintering mechanism involves creation of vacancies in the curved surfaces of pores, their transport through the grain and their absorption at grain boundaries, which play the role of sinks, Fig. 3.11. Material transport in ionic compounds requires the simultaneous flux of both cations and anions to preserve both electro-neutrality and stoichiometry. The theory of sintering for ionic solids is more complex because of the constraints imposed by the electrical charges of the diffusing species. A difference in mobility between diffusing ions, for instance, gives rise to electric fields opposing the flux of fast species. Material transport is therefore controlled by the slowest diffusing species.

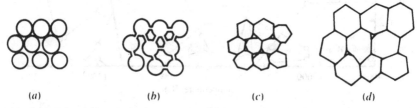

(a)   (b)   (c)   (d)

Fig. 3.16. Schematic representation of sintering stages: (a) greenbody, (b) initial stage, (c) intermediate stage, and (d) final stage.

Dissolution of small amounts of aliovalent compounds in ionic materials can produce an 'extrinsic', or additional, concentration of point defects (with respect to that in thermodynamic equilibrium), which may lead to enhanced volume diffusion. In the simple case of sintering of an ionic solid MeO, with Me = divalent cation, dissolution of a monovalent oxide L_2O would have the tendency to create anion vacancies or interstitial cations, while addition of a trivalent oxide A_2O_3 would increase the cation vacancy concentration. In oxides, the slowest diffusing species is the O^{2-} ion; creation of anion vacancies would therefore lead to a higher densification rate. In spinels (Reijnen, 1967), as well as in hexagonal ferrites (Van den Broek & Stuijts, 1977), higher final densities are effectively obtained by addition of a small excess of divalent oxide. Assuming that the concentration of interstitial cations is negligible, Reijnen (1970) has calculated that the condition for maximum volume diffusion occurs for $D_oC_o = D_cC_c$, where the diffusion constants for anion and cation are D_o and D_c, respectively, and C_o and C_c represent the anion and cation vacancy concentrations, respectively. Since in most oxides the cation diffusion constant is larger than the oxide diffusion constant, the maximum sintering rate is obtained for $C_o > C_c$.

A pore in a fine-grain matrix experiences a compressive stress due to the formation of grain boundary area and the disappearance of surface area during shrinkage. The internal pressure of a gas trapped in the pore resists its elimination. The decrease in pore volume continues until the compressive stress equals the internal pressure (Lin & German, 1988). The shrinkage rate is thus affected by the pore size (Evans & Hsueh, 1986), and the ratio of pore size to grain size (De Jonghe & Rahaman, 1984).

The presence of insoluble impurities may produce local differences in densification rate, which result in mechanical stresses. Sintering damage such as cracks and arrays of pores are associated with the elimination of these stresses (Hsueh, Evans & McMeeking, 1986).

Grain growth begins during the intermediate stage of sintering. Since grain boundaries are the sinks for vacancies, grain growth tends to decrease the pore elimination rate due to the increase in distance between pores and grain boundaries, and by decreasing the total grain boundary surface area.

In the final stage, grain growth is considerably enhanced and the remaining pores may become isolated. When the grain growth rate is very high, pores may be left behind by rapidly moving grain boundaries, resulting in pores that are trapped inside the grains, and not between the grains. This *intragranular* porosity, Fig. 3.17, is practically impossible to eliminate, leading to poor magnetic and mechanical properties. Exaggerated or *discontinuous grain growth* is characterised by the excessive growth of some grains at the expense of small, neighbouring ones, trapping all the pores present in that volume, Fig. 3.18. Discontinuous growth is believed to result from one or several of the following: powder mixtures with impurities; a very large distribution of initial particle size; sintering at excessively high temperatures; in ferrites containing Zn and/or Mn, a low O_2 partial pressure in the sintering atmosphere.

Grain growth kinetics depend strongly on impurity content. A minor dopant can drastically change the nature and concentration of defects in the matrix, affecting grain boundary motion, pore mobility and pore removal (Yan & Johnson, 1978). The effect of a given dopant depends on its valence and solubility with respect to the host material. If it is not soluble at the sintering temperature, the dopant becomes a second phase which usually segregates to the grain boundary. Its mobility is very small and represents a drag on grain boundary motion. However, if the sintering temperature is very high, a grain boundary can break away from impurities and produce exaggerated grain growth. If the dopant is soluble in the host material at the sintering temperature, it can increase the transport rate of the slow moving species, by increasing the appropriate defect concentration. On cooling, the solubility of dopants generally decreases and they may segregate to the grain boundaries. Finally, grain growth can be slowed by the presence of a liquid phase. Even a very small amount of liquid in the grain boundaries reduces the driving force for grain growth. In some cases, this is an effective method for attaining a high final density with small grains.

The atmosphere during sintering is particularly important. The O_2 partial pressure in the gas phase in equilibrium with Fe^{3+} is a function

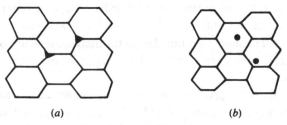

(a)   (b)

Fig. 3.17. Porosity character: (a) intergranular, (b) intragranular.

of the ferrite composition and temperature. A low O_2 partial pressure results in the reduction $Fe^{3+} \rightarrow Fe^{2+}$, and the tendency to increase O vacancies in the solid. This process enhances volume diffusion, but is difficult to control and leads to intragranular porosity. Also, the presence of Fe^{2+} in the final ferrite microstructure results in a high electrical conductivity, with high losses at elevated frequency. Ferrites containing Mn, which have many important applications, have an additional complication in their preparation technology, since this cation also has two valence states separated by a low energy difference. Slick (1971) has determined a composition–atmosphere–temperature diagram for the phase equilibrium in Mn–Zn ferrites, Fig. 3.19.

High sintering temperatures can lead to the partial volatilisation of some cations, particularly Zn and Cd. Zn volatilisation has been observed

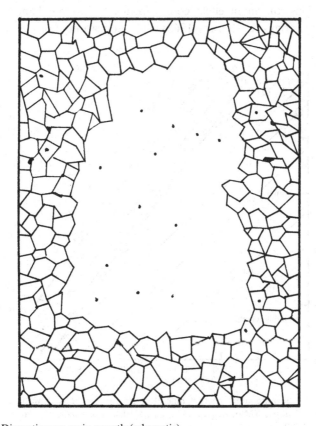

Fig. 3.18. Discontinuous grain growth (schematic).

to depend also on the O_2 partial pressure in the gas phase (Sainamthip & Amarakoon, 1988). Zn volatilisation results in increased grain growth.

3.4.3   Hot press sintering

Application of pressure during sintering results in a lower final porosity as compared with pressureless processes. External pressure increases the pressure difference created by curved surfaces, Eq. (3.6), and facilitates pore elimination. For the usual pressures in hot press sintering (or pressure-assisted sintering), this increase in pressure difference is 5–10 times (McColm & Clark, 1988). As a result, sintering temperatures can be reduced by $\sim 300\,^\circ\mathrm{C}$ and higher final densities are obtained.

Other mechanisms, such as plastic flow and particle rearrangement can contribute to densification under pressure; experimental results, however, indicate that the main effect comes from a pressure-enhanced volume diffusion (Spriggs & Atteraas, 1968). The application of pressure has little

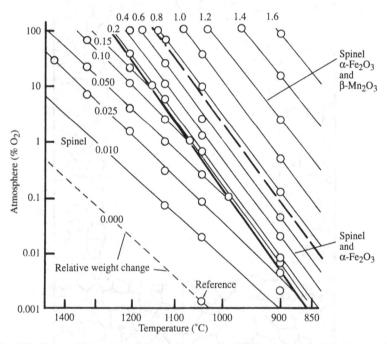

Fig. 3.19. Composition–atmosphere–temperature diagram for the ferrite with initial overall composition $(ZnO)_{18.3}(MnO)_{26.8}(Fe_2O_3)_{54.9}$ at 1050 °C. Lines indicate relative weight change (Slick, 1971).

effect on grain growth; it is therefore possible to obtain a fully densified material at temperatures at which grain growth is negligible.

A variation in this method is *hot isostatic pressing* (HIP), where the compaction is performed in the isostatic mode. HIP has been used for ferrites in applications where a high density and small grain size are crucial, such as microwave materials and magnetic recording heads. A method for HIP with no container has been developed (Takama & Ito, 1979). The presintered material is fired a second time at 1200–1300 °C in a gas pressure of 100 MPa. As no encapsulation is used, contamination of the external surface is avoided. An important requirement for the successful application of this method is that the presintered material posseses only closed porosity. The reason is that in the case of open porosity the increase of pressure in the solid is the same as in the pores which eliminates the pressure gradient to suppress pores. In the case of closed porosity, in contrast, the external pressure can eventually overcome the internal pressure of the closed pores. In fact, open porosity can result in extensive surface damage during HIP (Kellet & Lange, 1988).

3.4.4 Microwave sintering

Electromagnetic fields can penetrate and propagate through many materials. The interaction of these fields with free and bound charges, as well as with magnetic moments leads to an energy transfer capable of performing high-temperature treatments, or *microwave sintering*. Since Curie points for all the ferrites are below the typical sintering range, their microwave heating characteristics are dominated only by the dielectric loss spectrum. Ba hexaferrite particles have been sintered by microwave techniques (Krage, 1981), at 1230 °C with a microwave signal of 2450 MHz. High-quality ferrites were obtained; the maximum heating rate without cracking of the samples was 9 °C/min. Properties of the final materials were comparable with those of conventionally sintered ferrites.

An alternative method which permits the microwave sintering of low-electrical-conductivity materials is *secondary* microwave processing. In this method, Fig. 3.20, a highly efficient microwave absorber (a *susceptor*) is used to heat a cavity where the sample being processed is placed (Pope, 1991). This method should allow microwave processing of a wide range of ferrites.

3.5   *Permanent magnet technology*

The science and technology of permanent magnets encompasses basic theories of magnetisation mechanisms as well as more practical details of fabrication methods. The performance of a given material as a permanent magnet, or *hard magnet*, depends critically on its microstructure. For a given ferrite composition, remanent magnetisation and coercive field, for instance (see next chapter), are determined by grain size and the degree of preferred orientation of the *c*-axis. To a considerable extent, the technology of ferrite permanent magnets focuses on these two micro-structural aspects.

Barium hexaferrite, $BaFe_{12}O_{19}$, was developed in 1952 by Philips Industries (Netherlands) under the registered name of *Ferroxdure*. Strontium hexaferrite, $SrFe_{12}O_{19}$, with improved properties as compared with barium ferrite, was introduced in the 1960s, also by Philips. Other hexagonal ferrites such as $BaFe_2^{2+}Fe_{16}^{3+}O_{27}$ (Lotgering, Vromans & Huyberts, 1980) and $BaZn_2Fe_{12}O_{27}$, $BaZnFe_{14}O_{23}$ and $Ba_2ZnFe_{18}O_{23}$ (Kools, 1991) possess comparable or superior properties to Ba or Sr ferrites. However, large-scale commercial permanent ferrite magnets are dominated by Sr and Ba ferrites, as a result of their well-known technology

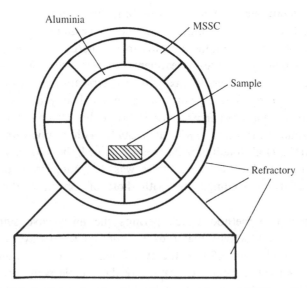

Fig. 3.20. Schematic furnace for secondary microwave sintering: MSSC = microwave susceptor ceramic composite (Pope, 1991).

and low prices. Permanent magnets prepared from metallic alloys (see Chapter 6) offer properties clearly superior to those of ferrites and are used in specialised applications; however, ferrites retain a considerable share of the world consumption because of their low production cost.

Typical processing routes for ferrite permanent magnets are shown schematically in Fig. 3.21. A very wide variety of products are obtained for many different applications, ranging from *sintered anisotropic*, the strongest ferrite permanent magnet, to the *plastoferrites*, which are made of ferrite particles embedded in a flexible (plastic or rubber) matrix. Plastoferrites are used in applications where a flexible magnet is needed. An overview of the processing routes for these two products is given below.

Powder preparation is performed by mixing inexpensive mineral raw materials, $BaCO_3$ or $SrCO_3$ and Fe_2O_3. Except for one of the products (*sintered isotropic*, Fig. 3.21), powder preparation involves a prefiring at 1250–1300 °C in continuous rotary kilns, leading to the ferrite (Van den Broek and Stuijts, 1977). Milling is the next step, which is accomplished in wet or dry ballmills, depending on the product. Final particle size and size distribution are critical for the ferrite coercive field, since this property shows a maximum as a function of milling time (Van den Broek and Stuijts, 1977), corresponding to a particle size ~ 1 μm. Ferrite powder in the plastoferrites route is mixed with plastics or rubber before shaping.

One of the most important steps in permanent magnet technology is *shaping* greenbodies with strong preferred orientation of particles. This is accomplished by pressing a concentrated suspension of fine ferrite particles ($\sim 15\%$ water) in a strong magnetic field (~ 10 kOe). Plastoferrites are shaped by *injection moulding* or *rolling* also in a magnetic field, before the plastic matrix settles.

Sintering of ferrite permanent magnets is difficult because it requires full densification without grain growth. Addition of silica, SiO_2, has been shown to facilitate densification without grain growth by liquid-phase sintering (Reijnen, 1991). Silica is insoluble in the ferrite phase, segregates to grain boundaries and hinders grain boundary motion (Den Broeder & Franken, 1981). The preferred orientation is not lost during sintering; it is enhanced by the controlled growth of well-aligned grains at the expense of grains with other orientations.

Finishing of sintered anisotropic ferrites involves grinding surfaces to specified dimensions. Since sintered ferrites are hard and brittle, grinding is usually performed with diamond tools. A serious problem in sintering hexaferrites is anisotropic shrinkage. It can be as high as 25% parallel,

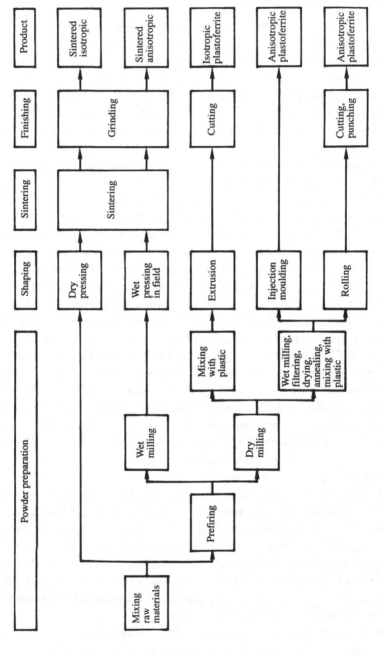

Fig. 3.21. Processing routes for ferrite permanent magnets. (Adapted from Van den Broek & Stuijs, 1977.)

and 10% perpendicular, to the preferred orientation, respectively. These differences have to be taken into account in the design of green pieces and pressing dies. The thermal expansion coefficient is also anisotropic, and can lead to fracture during cooling. Finishing plastoferrites consists of cutting (or punching) the desired shapes out of rolled laminates.

3.6   Preparation of ferrite thin films

Many specialised applications of magnetic materials involve the utilisation of ferrite thin films, such as magnetic and magnetooptic recording media, microwave devices in integrated circuits and coatings for microwave shielding. To prepare these ferrite thin films, a wide variety of techniques has been devised.

3.6.1   Liquid phase epitaxy

Liquid phase epitaxy, or *LPE*, has been investigated in great detail because this was the first method for preparing monocrystalline garnet films for use as the ferrites in 'bubble' computer memories, and more recently, for applications in magnetooptics.

The term *epitaxy* is used to describe the growth of a monocrystalline film of one material in a definite crystallographic orientation on a crystal face of another material, the *substrate*. To promote growth, there must be a correspondence of both lattices at the interface. A misfit between film and substrate results in stresses as the film grows laterally, which may affect the stability of the deposited crystal. The magnetic properties of LPE garnet films can be optimised by varying this misfit.

Substrates are non-magnetic solid solutions of rare-earth garnets with Ca^{2+}, Mg^{2+} and Zr^{2+} ions. By varying the composition, almost any lattice parameter in the range 12.290–12.620 Å can be obtained (Mateika, Laurien & Rusche, 1982).

To prepare a garnet film by LPE, a melt is obtained with the rare-earth (or Y) and Fe oxides in the appropriate proportions, in a PbO/B_2O_3 flux, at typical temperatures of 970–1100 °C, Fig. 3.22. A small excess of Fe_2O_3 is added to maintain the composition within the $Fe_2O_3-[PbO/B_2O_3]-Y_3Fe_5O_{12}$ triangle, with the garnet as the primary phase field. A large supercooled state (10–40 °C) can be produced in the melt; the substrate is placed just above the melt surface for a few minutes to ensure that its temperature is the same as that of the melt; it is then dipped into the melt. Growth takes place because the melt is in a supersaturated state. The substrate is usually disc-shaped and held in a horizontal position to allow rotation on its axis and promote homogeneous growth. After a few minutes, it is extracted from the melt. Growth takes place on the (111) face of the substrate, with the same orientation for the film. Typically, the thickness of the film is 1–4 µm. The cooling rate is very slow, to avoid temperature gradients between the film and the substrate that could affect film quality.

A model describing growth kinetics in LPE has been proposed by Van Erk (1978). The garnet solubility was described as the solubility product of the individual ions (rare-earth and iron) in a three-component system: rare-earth, Fe and solvent (PbO/B_2O_3). The rate-limiting steps in the total process were the diffusion through the boundary layer and the incorporation of the rare-earth species in the solid phase. The composition of the film was thus growth-rate-dependent, as observed experimentally (White & Wood, 1972; Tolskdorf, Bartels & Tolle, 1981).

Single-crystal spinel ferrite films can be epitaxially grown on MgO substrates. The ferrite, in a melt with Na_2CO_3 as solvent, is deposited on the substrate at temperatures of 1100–1250 °C. Crystallisation of the

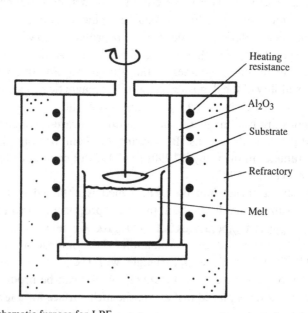

Fig. 3.22. Schematic furnace for LPE.

ferrite film occurs by solvent evaporation (Gambino, 1967). The best results were obtained by spreading dry anhydrous Na_2CO_3 on the substrate before covering it with the dry ferrite powder.

The use of a laser source under O_2 pressure can be advantageous to produce changes in the magnetic properties of LPE garnets without affecting film stability (Herman, DeLuca & Vollmer, 1981). In this technique, also known as *laser annealing*, high temperatures were produced in small, localised film areas with high heating and cooling rates. An O_2 partial pressure of 5 atm in the gas phase allowed annealings at 1570 °C without O loss in the film. Laser annealing could therefore be used to produce and quench a high-temperature cation distribution.

3.6.2 Sputtering techniques

An incident ion beam on the surface of a solid can produce many interactions, depending mainly on the kinetic energy of the ions. When this energy is in the range 10^2–10^3 eV, atoms are ejected from the solid surface. The process can be controlled in such a way that the atoms can be 'peeled off' from the solid, known as the *target*, and redeposited also in a controlled way onto another surface, or *substrate*. This is an effective momentum transfer and not an evaporation phenomenon. The process is known as *sputtering*; the main controlling factors are the energy of the bombarding ions, their molecular weight, the nature and conditions of the target surface and the pressure in the space between the target and the source of ions. The ions are typically obtained from a gas discharge forming a plasma, between a metallic anode and the target, which plays the role of the cathode, Fig. 3.23. This is a general technique for thin film preparation.

A number of variations have been introduced to improve the efficiency of the deposition process. Instead of a dc potential, a radio-frequency voltage can be used to maintain the plasma ('rf-sputtering'). Deposition rates may be increased by adding a focusing system or *magnetron*. Another improvement makes use of a separate anode to produce the plasma and to preserve the quality of the target. This system is known as a *triode*.

One of the great successes of sputtering techniques is the preparation of monocrystalline thin films of Bi–Fe garnet, $Bi_3Fe_5O_{12}$ (or *BiIG*) (Gomi, Satoh & Abe, 1989; Okuda *et al.*, 1990), by *reaction ion beam sputtering*, or *RIBS*. In this variation of sputtering, the presence of an additional component in the gas phase (usually oxygen), produces a chemical reaction during the film deposition, Fig. 3.24. The substrate is kept at a relatively high temperature (300–500 °C) to promote a better crystallisation of the garnet. Deposition rates are typically 3–7 nm/min, and thicknesses, typically 600 Å–1 µm. The target composition is the same as the film and

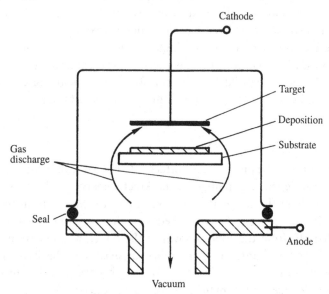

Fig. 3.23. Basic sputtering apparatus.

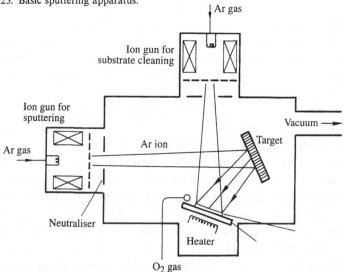

Fig. 3.24. Schematic device for RIBS. (Adapted from Okuda *et al.*, 1990.)

the substrate is $NdGa_5O_{12}$ ('NGG'). This garnet is also prepared as an amorphous thin film by sputtering on glass substrates at 400 °C and a postdeposition annealing in air at 650 °C to transform it into a polycrystalline film (Hirano, Namikawa & Yamazaki, 1991).

Magnetite thin films have also been prepared by reactive sputtering (Ortiz *et al.*, 1988). Fe was used as the target and the substrate was Si single crystals with (100) orientation, with O_2 as the reactive gas. Polycrystalline films of magnetite were obtained with a thickness of 1000 Å. The films obtained were further oxidised to γ-Fe_2O_3 in a conventional furnace at 275–300 °C.

Another important ferrite with thin film applications is $BaFe_{12}O_{19}$. Single-crystal films can be prepared by LPE using a well-textured $BaFe_{12}O_{19}$ sputtered film as a seeding layer (Yuan, Glass & Adkins, 1988). The best substrate for this sputtering process was single-crystal Gd–Ga garnet $Gd_3Ga_5O_{12}$ ('GGG') (Lacroix *et al.*, 1991).

3.6.3   Other methods

Many other methods have been devised to prepare ferrite thin films. Zn- and Al-substituted cobalt ferrites have been prepared by vacuum evaporation onto fused quartz substrates (Suzuki, Namikawa & Yamazaki, 1988), followed by a heat treatment. Polycrystalline Ni–Zn–Co ferrite thin films have been prepared by pyrolitic hydrolysis of metal inorganic salts (Gleason & Watson, 1963). The solution of the desired cations was atomised on glass substrates heated at 800 °C. Garnet ferrites can also be prepared by pyrolysis of the corresponding nitrates on glass substrates. The deposition is improved by the *spray-spin* coating technique, Fig. 3.25, which consists of atomising the solution on a rapidly rotating substrate (~ 400 rpm) (Cho, Gomi & Abe, 1991). To decompose the metal nitrates, films were preheated at 400 °C and then annealed at 680 °C for several hours. The thickness was about 0.5 µm. Larger grains were obtained by first preparing a thin, fine-grained layer which was annealed at 550–680 °C and then depositing a second layer on it. Nucleation of grains of the second layer was facilitated by the first layer. A uniform grain size has been obtained by successive depositions of 300–600 Å layers (Mizuno & Gomi, 1986).

Sol–gel techniques have also been used for film preparation. Magnetite has been prepared from sols formed from tris(acetylacetonate, $CH_3COCHCOCH_3$)FeIII, or $Fe(acac)_3$ and a mixture of CH_3COOH as solvent and HNO_3 as catalyst (Tanaka *et al.*, 1989). The sol was applied to a silica glass substrate by dipping the substrate into the sol, and pulling it out at a constant speed (0.6 mm/s). The film was then annealed at 940 °C, 10 min in air, and the cycle repeated ten times, to obtain a 0.2 μm film of α-Fe_2O_3. Magnetite was obtained by a heat treatment at 500–700 °C in air, with the film embedded in carbon. The steps of the process are shown schematically in Fig. 3.26.

Chemical vapour deposition (CVD), or more specifically, organo-metallic CVD can also be used to prepare ferrite thin films (Itoh, Takeda & Naka, 1986). A mixture of acetylacetonate complexes of the desired metal is evaporated in a quartz boat (number 1 in Fig. 3.27) and is reacted with O_2, which is introduced directly in the deposition zone. By adding a second furnace (B), an additional component evaporating at a different temperature can be transported to the substrate. Glass or MgO single crystals can be used as the substrates; an annealing up to 1000 °C is performed on the film to improve its crystallisation. Spinel thin films (Ni, Ni–Zn) have been prepared by this method.

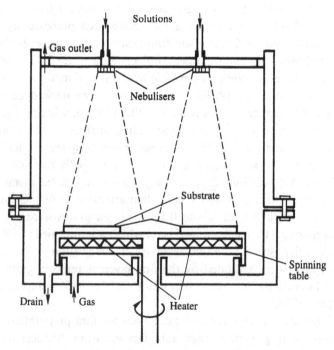

Fig. 3.25. Apparatus for the spray-spin coating technique. (Adapted from Abe *et al.*, 1987.)

Preparation of ferrite thin films

Plating processes allow the deposition of thin films at temperatures as low as 70 °C on low-heat-resistant substrates such as GaAs. Integrated circuits, as well as multilayers with organic layers alternating with ferrite layers, can therefore be prepared by this technique (Abe, *et al.*, 1988). Plating is performed by depositing fine droplets (*nebulising*) of a $FeCl_3$–CH_3COONH_4 solution, Fig. 3.25, onto a heated, rotating (400 rpm) substrate ('spray-spin' coating). The film is oxidised to Fe_3O_4 by air in the deposition chamber, or by adding a solution of $NaNO_3$, $NaNO_2$ or H_2O_2 to the metal solution (Abe *et al.*, 1987). Deposition on Al substrates is improved by a previous coating of SiO_2 exposed to an air plasma. Grain growth occurs in columnar mode, with a final grain size ~ 1000 Å, and a typical thickness of ~ 3000 Å. The films are oxidised to γ-Fe_2O_3 by annealing at 300 °C in air.

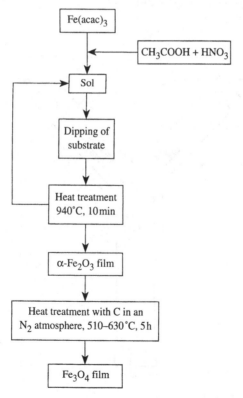

Fig. 3.26. Schematic sol–gel process for the preparation of ferrite thin films. (Adapted from Tanaka *et al.*, 1989.)

Plating has also been done using thin liquid–film techniques; aqueous solutions were simply in contact with the substrate (Itoh, Hori & Abe, 1991). The reaction solution was $FeCl_2$ + MeCl and ammonium acetate, CH_3COOCH_4, where Me = Ni, Mn, Co or Ni–Zn. An oxidising solution ($NaNO_3$ + CH_3COOCH_4) was added to obtain the corresponding ferrite. The substrate was kept at a low temperature ($\sim 60\,°C$). It has been shown that the irradiation of the substrate surface by means of a xenon lamp ($450\ W/m^2$) increased the deposition rate by 5–10 times, to 150–320 nm/min; this is known as *light-enhanced ferrite plating*, Fig. 3.28.

Arc plasma has been used to deposit polycrystalline Ni–Zn, Mg–Mn spinels and garnets onto diverse substrates (Harris *et al.*, 1970). The arc gases are passed between concentric electrodes and rapidly heated by a dc arc, Fig. 3.29. The material to be deposited is injected as a fine powder with a carrier gas into the plasma. The film thickness is typically 1 μm with deposition rates of 2 μm/min.

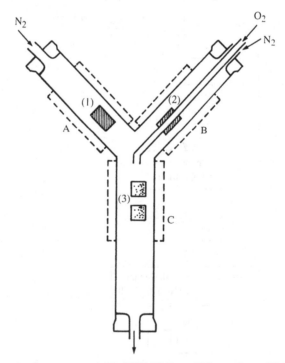

Fig. 3.27. Apparatus for organo-metallic CVD. (1) Fe and Ni complexes; (2) Zn complex; (3) substrate. A, B and C: furnaces. (Adapted from Itoh *et al.*, 1986.)

Preparation of ferrite thin films

Plasmas are also used to provide a more reactive O_2 atmosphere during molecular beam epitaxy of magnetite (Lind *et al.*, 1991). In this technique, electrons are accelerated to impinge on a metallic target (Fe) to release metal atoms and deposit them on a substrate (the (001) face of MgO single crystals). Instead of flooding the deposition chamber with diatomic O_2 gas, an oxygen plasma provides a reactive mixture of O and O^+ which

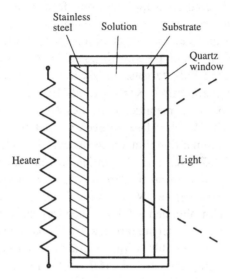

Fig. 3.28. Schematic arrangement for ferrite thin film preparation by light-enhanced plating. (Adapted from Itoh *et al.*, 1991.)

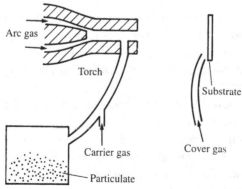

Fig. 3.29. Preparation of ferrite thin films using arc plasma. (Adapted from Harris *et al.*, 1970.)

87

oxidises the Fe to Fe_3O_4 and leads to a single crystal film on the substrate. The use of *plasma-assisted oxygen* thus allows film growth at low pressures. By using two metallic targets and an automatic shutter system, multilayered Fe_3O_4/NiO films with bilayer thicknesses as small as 17 Å can be obtained.

A method for the preparation of ceramic thin films, which is attracting attention, is laser evaporation, or *laser ablation*. In this technique, a laser of high intensity is used to evaporate atoms from a target and deposit them onto a substrate. One of the main advantages is that the localised, strong heating evaporates the target atoms as a whole; thin films of complex materials can be prepared with essentially the same composition as the corresponding bulk targets. Ni–Zn ferrite thin films have been prepared by laser evaporation (Konkar *et al.*, 1989). Two different types of targets were used: a metallic alloy $Fe_{50}Ni_{50}$ and a ceramically synthesised Zn ferrite. A pulsed laser source was used; typically, 400 pulses were required to obtain a thin film, on substrates of Al_2O_3 single crystals and quartz, at 673–773 K. The O_2 partial pressure during deposition determined the final state of the film; for $p = 10^{-4}$ Pa, a metallic thin film with the alloy composition $Fe_{50}Ni_{50}$ was obtained; at $p = 6.7 \times 10^{-2}$ Pa, however, the thin film obtained exhibited the x-ray diffraction pattern of Ni ferrite. In the case of Zn ferrite target, a low O_2 partial pressure during deposition resulted in a thin film with a mixture of $Zn_xFe_{3-x}O_4$, $Fe_{1-x}O$ and ZnO. A higher O_2 partial pressure (6.7×10^{-2} Pa) led to two phases; the main phase was $Zn_{0.6}Fe_{2.4}O_4$.

3.7   Preparation of ferrite single crystals

Single crystals are essential for certain applications as well as for fundamental studies. Magnetic properties such as anisotropy and magnetostriction (see next chapter) are not the same for all the crystal orientations. Investigation of basic magnetic phenomena is also simplified if single crystals are available.

A simple method of preparing single crystals is to melt the material and to crystallise it by slow cooling; but practically all the ferrites dissociate at temperatures below their melting points, with loss of O_2. To overcome this problem several techniques have been used which are based on the formation of a flux, a solution, or a volatile chemical compound to obtain ferrite single crystals.

In the *Bridgman–Stockbarger* method, the material is melted in a conical crucible. After homogenisation, the temperature is slowly lowered, which leads to the nucleation of a seed in the conical tip of the crucible which then grows as a single crystal. Co ferrite ($CoFe_2O_4$) single crystals have been prepared by this technique, from a melt 95% ferrite and 5% $NaFeO_2$ (Ferreti, Kunnmann & Wold, 1963). The initial cobalt ferrite was prepared by the pyridinate precursor method to avoid the impurities often introduced by ballmilling or other ceramic techniques. $NaFeO_2$ is a good second component because it has liquid miscibility with Co ferrite, but does not enter into solid solution. The melt was obtained at 1590 °C under a pressure of 11.5 MPa to prevent the reduction of Fe^{3+} to Fe^{2+}. After controlled cooling and decreasing the pressure, the crucible was peeled off from the crystals. The $NaFeO_2$ was found at the top of the crucible, as this was the last part to solidify. It was separated from the Co crystals by diamond cutting. The largest crystals were ~ 2.5 cm in length and 0.8 cm in diameter.

Garnet single crystals have been obtained by the Bridgman–Stockbarger method; *BiCaVIG*, $Bi_{3-2x}Ca_{2x}V_xFe_{5-x}O_{12}$, with $0.96 \leq x \leq 1.35$, has been prepared from a flux formed by Bi_2O_3, $CaCO_3$, V_2O_5, Fe_2O_3 and PbO (Hodges *et al.*, 1967). In this case, Bi_2O_3 was part of the crystal as well as part of the flux.

The single crystal grown onto a rotating seed is pulled from the melt in the *Czochralski–Kryopoulos* method, Fig. 3.30. The flux is formed by the

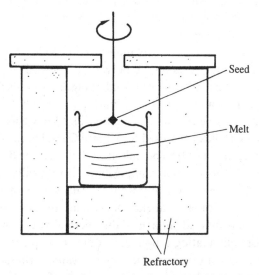

Fig. 3.30. Furnace for growing ferrite crystals by the Czochralski–Kryopoulos technique.

ferrite and other low melting oxides, such as $NaFeO_2$, BaO/B_2O_3 or BaO/Bi_2O_3. The liquidus temperature, which is the temperature at which the first solid appears in a supercooled solution, is determined simply by decreasing the temperature and observing the surface of the melt. Once determined, the crucible is kept at this temperature, and the seed is slowly rotated and withdrawn; typical values for rotation and withdrawal are 60–100 rpm and 0.025–0.075 mm/h, respectively. Ni ferrite (Kunnmann, Ferreti & Wold, 1963), YIG and Li ferrite, $LiFe_5O_8$ (Linares, 1964), and $Ba_2Zn_2Fe_{12}O_{22}$ ('Zn–Y') (AuCoin, Savage & Tauber, 1966) single crystals have been obtained by the Czochralski–Kryopoulos method.

The *hydrothermal method* involves dissolution of the ferrite in an alkaline medium, generally in NaOH, at high temperatures and pressures. The solution is formed in an autoclave, Fig. 3.31; in the lower part, the ferrite in the form of powder, or the corresponding oxides in stoichiometric proportions, form the *nutrient* zone. A baffle separates this region from the *growth* region, where several seeds are fixed to initiate the growth of crystals. The nutrient zone is kept at a slightly higher temperature than the growth zone; the ferrite components are dissolved in the nutrient zone and transported by convection to the growth zone, where they are deposited onto the seed in the cooler region. More ferrite will then dissolve in the nutrient zone and the cyclic process continues. Laudise & Kolb (1962) have obtained YIG single crystals by this technique, with nutrient and growth temperatures of 320 and 420 °C, respectively, at pressures in the 200–1350 atm range. The growth rate was 0.08 mm/day. Crystals with dimensions of about 1 cm were obtained.

A method similar to the zone-refining of semiconductors can be used to prepare ferrite single crystals. A polycrystalline rod of the ferrite is zone-melted; as the melted zone is moved along the rod, recrystallisation occurs. If a seed is added at the initial end, the whole rod can be transformed into a single crystal. The method, also known as the *travelling solvent floating zone*, has been used to prepare Ga-substituted YIG single crystals (Torii *et al.*, 1979). The initial rod was formed from oxides corresponding to the desired composition, isostatically pressed and sintered at 1600 °C in an O_2 atmosphere. The length of the rod was typically 120 mm, with a diameter of 12 mm; the solvent zone was ~ 10 mm. An important advantage of this method is that no crucible is needed, therefore eliminating a potential source of impurities.

A different approach is used in the *chemical vapour transport technique* (CVT). In this method, the solid is chemically transformed into a volatile compound, which is then transported in the gas phase. The chemical reaction is reversible, permitting the deposition of the solid onto a growing seed at a lower temperature, Fig. 3.32. This is different from physical transport, since the ferrite vapour pressure is negligible at

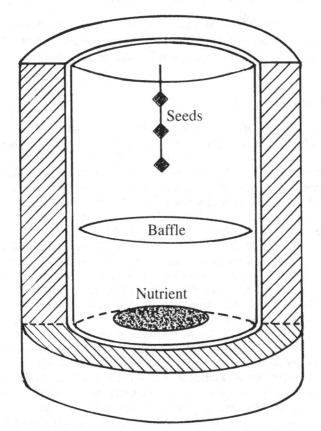

Fig. 3.31. Apparatus for growing ferrite crystals by the hydrothermal method. (Adapted from Rao & Gopalakrishnan, 1989.)

$T = 1000\,°C$ $T = 750\,°C$

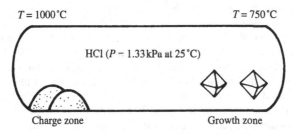

Charge zone   Growth zone

Fig. 3.32. Growing ferrite crystals by CVT. (Adapted from Wold, 1980.)

the temperatures involved. Kershaw & Wold (1968) have prepared magnetite single crystals by CVT in a HCl atmosphere, based on the reaction:

$$Fe_3O_4 + HCl \leftrightharpoons FeCl_2 + 2FeCl_3 + 4H_2O \tag{3.9}$$

Temperatures were 1000 and 750 °C in the *charge* and *growth* zones, respectively. The initial HCl pressure (at room temperature) was 1.33 kPa (10 mm Hg). The size of the crystals was difficult to control.

References

Abe, M., Tamaura, Y., Oishi, M., Saitoh, T., Itoh, T. & Gomi, M. (1987). Plating of ferrite film on 8″ disc at 70 °C by 'spray-spin-coating' method. *IEEE Transactions on Magnetics*, **MAG-23**, 3432–4.

Abe, M., Itoh, T., Tamaura, Y. & Gomi, M. (1988). Ferrite-organic multilayer film for microwave monolithic integrated circuits prepared by ferrite plating based on the spray-spin-coating method. *Journal of Applied Physics*, **63**, 3774–6.

Arendt, R. H. (1973). The molten salt synthesis of single domain $BaFe_{12}O_{19}$ and $SrFe_{12}O_{19}$ crystals. *Journal of Solid State Chemistry*, **8**, 339–47.

AuCoin, T. R., Savage, R. O. & Tauber, A. (1966). Growth of hexagonal ferrite crystals by a modified pulling technique. *Journal of Applied Physics*, **37**, 2908–9.

Brook, R. J. (1989). The materials science of ceramic interfaces. In *Surfaces and Interfaces of Ceramic Materials*, Vol. 173. Eds L. C. Dufour, C. Monty and G. Petot-Ervas, NATO ASI Series, Kluwer, Dordrecht, pp. xxii–xxxvii.

Brook, R. J. (1991). Sintering: an overview. *Concise Encyclopedia of Advanced Ceramic Materials*. Ed. R. J. Brook, Pergamon Press, Oxford, pp. 438–40.

Brooks, K. G. & Amarakoon, V. R. W. (1991). Sol–Gel coating of lithium zinc ferrite powders. *Journal of the American Ceramic Society*, **74**, 851–3.

Broussaud, D., Aboudaf, M., Perriat, P. & Rolland, J. L. (1989). Oxidation state/green strength relationships in Mn–Zn ferrite. In *Advances in Ferrites: Proceedings of the Fifth International Conference on Ferrites, India, 1989*, Vol. 1. Eds. C. M. Srivastava and M. J. Patni, Oxford & IBH Publishing Co. PVT Ltd, Bombay, pp. 75–81.

Carter, R. E. (1961). Mechanism and solid state reaction between magnesium oxide and aluminum oxide and between magnesium oxide and ferric oxide. *Journal of the American Ceramic Society*, **44**, 116–20.

Chen, C. J., Bridger, K., Winzer, S. R. & PaiVerneker, V. (1988). A novel low-temperature preparation of Ni–Zn ferrite and the properties of the ultrafine particles formed. *Journal of Applied Physics*, **63**, 3786–8.

Cho, J., Gomi, M. & Abe, M. (1991). Bi-substituted iron garnet films with fine grains prepared by pyrolysis. *Journal of Applied Physics*, **70**, 6301–3.

Date, S. K., Deshpande, C. E., Kulkarni, S. D. & Shrotri, J. J. (1989). Synthesis of ultrafine particles of strontium ferrite by chemical coprecipitation with sodium hydroxide. In *Advances in Ferrites: Proceedings of* the Fifth International Conference on Ferrites, India, 1989. Eds. C. M. Srivastava and M. K. Patni. Oxford & IBH Publishing Co. PVT Ltd, Bombay, India, pp. 55–60.

De Jonghe, L. C. & Rahaman, M. L. (1984). Pore shrinkage and sintering stress. *Journal of the American Ceramic Society*, 67, C-214–15.

Den Broeder, F. J. A. & Franken, P. E. C. (1981). The microstructure of sintered strontium hexaferrite with silica addition investigated by ESCA and TEM. In *Advances in Ceramics*, Vol. 1: *Grain Boundary Phenomena in Electronic Ceramics*. Eds L. M. Levinson and D. C. Hill. American Ceramic Society, Ohio, pp. 494–501.

Evans, A. G. & Hsueh, C. H. (1986). Behavior of large pores during sintering and hot isostatic pressing. *Journal of the American Ceramic Society*, 69, 444–8.

Fan, X. & Matijevic, E. (1988). Preparation of uniform colloidal strontium ferrite particles. *Journal of the American Ceramic Society*, 71, C-60–2.

Ferreti, A., Kunnmann, W. & Wold, A. (1963). Growth of ferrous-free cobalt ferrite single crystals. *Journal of Applied Physics*, 34, 388–9.

Gambino, R. J. (1967). Preparation of single-crystal ferrite films. *Journal of Applied Physics*, 38, 1129–32.

Ghate, B. B., Holmes, R. J. & Pass, C. E. (1982). Environmentally induced permeability degradation in ferrite powders. *American Ceramic Society Bulletin*, 61, 484–90.

Gleason, F. R. & Watson, L. R. (1963). Ferrite films prepared by the pyrohydrolitic deposition. *Journal of Applied Physics*, 34, 1217–18.

Gomi, M., Satoh, K. & Abe, M. (1989). New garnet films with giant Faraday rotation. In *Advances in Ferrites: Proceedings of the Fifth International Conference on Ferrites, India, 1989*, Vol. 2. Eds. C. M. Srivastava and M. J. Patni, Oxford & IBH Publishing Co. PVT Ltd, Bombay, pp. 919–24.

Haberey, F. (1987). Preparation of M- and W-type hexaferrite particles by the glass crystallization method on the basis of the pseudo-ternary system Fe_2O_3–BaO–B_2O_3. *IEEE Transactions on Magnetics*, **MAG-23**, 29–32.

Harris, D. H., Janowiecki, R. J., Semler, C. E., Willson, M. C. & Cheng, J. T. (1970). Polycrystalline ferrite films for microwave applications deposited by arc-plasma. *Journal of Applied Physics*, 41, 1348–9.

Herman, D. A., DeLuca, J. C. & Vollmer, H. J. (1981). Laser annealing of gallium-substituted garnet epitaxial films in pressurized oxygen. *IEEE Transactions on Magnetics*, **MAG-17**, 2562–4.

Hibst, H. (1982). Hexagonal ferrites from melts and aqueous solutions, magnetic recording materials. *Angewande Chemie, International Edition in English*, 21, 270–81.

Hirano, T., Namikawa, T. & Yamazaki, Y. (1991). Magneto-optical properties of Ca-substituted Bi-YIG sputtered films. *Journal of Applied Physics*, 70, 6292–4.

Hodges, L. R. Jr, Wilson, W. R., Rodrigue, G. P. & Harrison, G. R. (1967). Growth, magnetic and microwave characteristics of monocrystalline BiCaVIG. *Journal of Applied Physics*, 38, 1127–8.

Hsueh, C. H., Evans, A. G. & McMeeking, M. (1986). Influence of multiple heterogeneities in sintering rates. *Journal of the American Ceramic Society*, **69**, C-64-6.

Itoh, H., Takeda, T. & Naka, S. (1986). Preparation of nickel and Ni–Zn ferrite films by thermal decomposition of metal acetylacetonates. *Journal of Materials Science*, **21**, 3677–80.

Itoh, H., Hori, S. & Abe, M. (1991). Light-enhanced ferrite plating of $Fe_{3-x}M_xO_4$ (M = Ni, Zn, Co and Mn) films in an aqueous solution. *Journal of Applied Physics*, **69**, 5911–14.

Jain, G. C., Das, B. K. & Avtar, R. (1976). Calcination study of spray-dried manganese–zinc ferrite powder. *Indian Journal of Pure and Applied Physics*, **14**, 796–9.

Kaczmarek, W. A., Ninham, B. W. & Calka, A. (1991). Structure and magnetic properties of aerosol synthesized barium ferrite particles. *Journal of Applied Physics*, **70**, 5909–11.

Kang, S-J. L., Kim, K-H. & Yoon, D. N. (1991). Densification and shrinkage during liquid-phase sintering. *Journal of the American Ceramic Society*, **74**, 425–7.

Kellet, B. J. & Lange, F. F. (1988). Experiments on pore closure during hot isostatic pressing and forging. *Journal of the American Ceramic Society*, **71**, 7–12.

Kendall, K. (1984). Connection between structure and strength of porous solids. In *AIP Conference Proceedings, No 107: Physics and Chemistry of Porous Media 1983*. Eds D. L. Johnson and P. N. Sen. American Institute of Physics, New York, pp. 78–88.

Kershaw, R. & Wold, A. (1968). Single crystals of triiron tetraoxide. *Inorganic Syntheses*, **11**, 10–14.

Kingery, W. D., Bowen, H. K. & Uhlman, D. R. (1976). *Introduction to Ceramics*, 2nd edition, Wiley Interscience, New York, pp. 476–7.

Konkar, V. N., Nawathey, R., Chaudhari, S. M., Kanetkar, S. M., Date, S. K. & Ogale, S. B. (1989). Synthesis of ferrite films by pulsed laser evaporation process. In *Advances in Ferrites: Proceedings of the Fifth International Conference on Ferrites, India, 1989*, Vol. 1. Eds. C. M. Srivastava and M. J. Patni. Oxford & IBH Publishing Co. PVT Ltd, Bombay, pp. 525–31.

Kools, F. (1991). Hard magnetic ferrites. In *Concise Encyclopedia of Advanced Ceramic Materials*. Ed. R. J. Brook. Pergamon Press, Oxford, pp. 200–6.

Kooy, C. (1965). Material transport in solid state reactions. In *Fifth International Symposium on the Reactivity of Solids, Munich*. Ed. G. M. Schwab. Elsevier, London, pp. 21–8.

Krage, M. K. (1981). Microwave sintering of ferrites. *American Ceramic Society Bulletin*, **60**, 1232–4.

Kunnmann, W., Ferreti, A. & Wold, A. (1963). Flux growth of $NiFe_2O_4$ crystals by the Czochralski method. *Journal of Applied Physics*, **34**, 1264.

Lacroix, E., Gerard, P., Marest, G. & Dupuy, M. (1991). Substrate effects on the crystalline orientation of barium hexaferrite films. *Journal of Applied Physics*, **69**, 4770–2.

Laudise, R. A. & Kolb, E. D. (1962). Hydrothermal crystallisation of yttrium-iron garnet on a seed. *Journal of the American Ceramic Society*, **45**, 51–3.

Lin, S. T. & German, R. M. (1988). Compressive stress for large-pore removal in sintering. *Journal of the American Ceramic Society*, **71**, C-432–3.

Linares, R. C. (1964). Growth of single-crystal garnets by a modified pulling technique. *Journal of Applied Physics*, **35**, 433–4.

Lind, D. M., Berry S. D., Chern, G., Mathias, H. & Testardi, L. R. (1991). Characterization of the structural and magnetic ordering of Fe_3O_4/NiO superlattices grown by oxygen-plasma-assisted molecular-beam epitaxy. *Journal of Applied Physics*, **70**, 6218–20.

Lotgering, F. K., Vromans, P. H. G. M. & Huyberts, M. A. H. (1980). Permanent-magnet material obtained by sintering the hexagonal ferrite $W = BaFe_{18}O_{27}$. *Journal of Applied Physics*, **51**, 5913–18.

Mateika, D., Laurien, R. & Rusche, Ch. (1982). Lattice parameters and distribution coefficients as function of Ca, Mg and Zr concentrations in Czochralski grown rare earth gallium garnets. *Journal of Crystal Growth*, **56**, 677–89.

Matsumoto, K., Yamaguchi, K. & Fujii, T. (1991). Preparation of bismuth-substituted yttrium iron garnet powders by the citrate gel process. *Journal of Applied Physics*, **69**, 5918–20.

Matsumoto, K., Yamanobe, Y., Sasaki, S. & Fujii, T. (1991). Preparation of YIG fine particles by mist pyrolysis. *Journal of Applied Physics*, **70**, 5912–14.

McColm, I. J. & Clark, N. J. (1988). *Forming, Shaping and Working of High Performance Ceramics*. Blackie, Glasgow, pp. 1–338.

Mizuno, T. & Gomi, M. (1986). Magneto-optical properties of bi-substituted garnet films prepared by pyrolisis. *IEEE Transaction on Magnetics*, **MAG-22**, 1236–8.

Okuda, T., Katayama, T., Kobayashi, H., Kobayashi, N., Satoh, K. & Yamamoto, H. (1990). Magnetic properties of $Bi_3Fe_5O_{12}$ garnet. *Journal of Applied Physics*, **67**, 4944–6.

Ortiz, C., Lim, G., Chen, M. M. & Castillo, G. (1988). Physical properties of spinel iron oxide thin films. *Journal of Materials Research*, **3**, 344–50.

Pope, E. J. A. (1991). Microwave sintering of sol–gel derived silica glass. *American Ceramic Society Bulletin*, **70**, 1777–8.

Qian, X. & Evans, B. J. (1981). Solid state and magneto-chemistry of the $SrO–Fe_2O_3$ system. IV. Synthesis of $SrFe_{12}O_{19}$ from coprecipitated precursors. *Journal of Physics*, **52**, 2523–5.

Rao, C. N. R. & Gopalakrishnan, J. (1989). *New Directions in Solid State Chemistry*. Cambridge University Press, Cambridge, p. 143.

Reijnen, P. J. L. (1967). Sintering behaviour and microstructure of aluminates and ferrites with spinel structure with regard to deviations from stoichiometry. *Science of Ceramics*, **4**, 169–88.

Reijnen, P. J. L. (1970). Nonstoichiometry and sintering in ionic solids. In *Problems of Nonstoichiometry*. Ed. A. Rabenau. North-Holland, Amsterdam, pp. 219–38.

Reijnen, P. L. J. (1991). Solid-state reactions. In *Concise Encyclopedia of*

Advanced Ceramic Materials. Ed. R. J. Brook. Pergamon Press, Oxford, pp. 445–54.

Ruthner, M. J. (1977). Fast reaction sintering process for the production of ferrites. *Journal de Physique*, **C1-38**, C1-311–15.

Sainamthip, P. & Amarakoon, V. R. W. (1988). Role of zinc volatilization on the microstructure development of manganese-zinc ferrites. *Journal of the American Ceramic Society*, **71**, 644–8.

Schnettler, F. J. & Johnson, D. W. (1971). Synthesized microstructure. In *Ferrites: Proceedings of the International Conference, July 1970, Japan.* Eds Y. Hoshino, S. Iida and M. Sugimoto. University of Tokyo Press, Tokyo, pp. 121–4.

Slick, P. I. (1971). A thermogravimetric study of the equilibrium relations between a Mn–Zn ferrite and an O_2–N_2 atmosphere. In *Ferrites: Proceedings of the International Conference, July 1970, Japan.* Eds. Y. Hoshino, S. Iida and M. Sugimoto. University of Tokyo Press, Tokyo, pp. 81–3.

Spriggs, R. M. & Atteraas, L. (1968). Densification of single phase systems under pressure. In *Ceramic Microstructures: Their Analysis, Significance and Production, Proceedings of the Third International Materials Symposium, 1966.* Eds R. M. Fulrath and J. A. Pask, Jr. Wiley, New York, pp. 701–27.

Srinivasan, T. T., Ravindranathan, P., Cross, L. E., Roy, R., Newnham, R., Sankar, S. G. & Patil, K. C. (1988). Studies on high density nickel zinc ferrites and its magnetic properties using novel hydrazine precursors. *Journal of Applied Physics*, **63**, 3789–91.

Stäblein, H. & Willbrand, J. (1972). Texture and reaction mechanism in the formation of barium hexaferrite. In *Proceedings of the Seventh International Symposium on the Reactivity of Solids, Bristol, July 1972.* Eds J. S. Anderson, M. W. Roberts and F. S. Stone. Chapman & Hall, London, pp. 589–97.

Suresh, K., Kumar, N. R. S. & Patil, K. C. (1991). A novel combustion synthesis of spinel ferrites, orthoferrites and garnets. *Advanced Materials*, **3**, 148–50.

Suzuki, K., Namikawa, T. & Yamazaki, Y. (1988). Preparation of zinc- and aluminium-substituted cobalt-ferrite thin films and their Faraday rotation. *Japanese Journal of Applied Physics*, **27**, 361–5.

Takada, T. (1982). Development and application of synthesising technique of spinel ferrites by the wet method. In *Ferrites: Proceedings of the Third International Conference on Ferrites, 1980, Japan*, Eds H. Watanabe, S. Iida and M. Sugimoto. Center for Academic Publications, Tokyo, pp. 3–8..

Takada, T. & Kiyama, M. (1971). Preparation of ferrites by wet method. In *Ferrites: Proceedings of the International Conference, July 1970, Japan.* Eds Y. Hoshino, S. Iida and M. Sugimoto. University of Tokyo Press, Tokyo, pp. 69–71.

Takama, E. & Ito, M. (1979). New Mn–Zn ferrites fabricated by hot isostatic pressing. *IEEE Transactions on Magnetics*, **MAG-15**, 1856–60.

Tanaka, K., Yoko, T., Atarashi, M. & Kamiya, K. (1989). Preparation of Fe_3O_4 thin films by the sol–gel method and its magnetic properties. *Journal of Materials Science Letters*, **8**, 83–5.

References

Tolskdorf, W., Bartels, G. & Tolle, H. J. (1981). Compositional inhomogeneities along the growth direction of substituted yttrium iron garnet epilayers. *Journal of Crystal Growth*, **52**, 722–8.

Torii, M., Kihara, U., Goto, H., Kimura, S. & Shindo, I. (1979). Homogeneity of YGaIG single crystal grown by the FZ method. *IEEE Transactions on Magnetics*, **MAG-15**, 1732–4.

Tseng, T. Y. & Lin, J. C. (1989). Preparation of fine grained Ni–Zn ferrites. *Journal of Materials Science Letters*, **8**, 261–2.

Van den Broek, C. A. M. & Stuijts, A. L. (1977). Ferroxdure. *Philips Technical Review*, **37**, 157–75.

Van Erk, W. (1978). The growth kinetics of garnet liquid phase epitaxy using horizontal dipping. *Journal of Crystal Growth*, **43**, 446–56.

Venturini, E. L., Morosin, B. & Graham, R. A. (1985). Magnetic and structural properties of shock-synthesized zinc ferrite. *Journal of Applied Physics*, **57**, 3814–16.

Wagner, C. (1936). Uber den Mechanismum der Bildung von Ionenverbindungen höherer Ordnung (Doppelsalze, Spinelle, Silikate), *Zeitung Physik und Chemie*, **B34**, 309–16.

White, E. A. D. & Wood, J. D. C. (1972). Heat and mass transfer in LPE processes. *Journal of Crystal Growth*, **17**, 315–21.

Wickham, D. G. (1967). Metal iron (III) oxides. *Inorganic Syntheses*, **9**, 152–6.

Wickham, D. G., Whipple, E. R. & Larson, E. G. (1960). The preparation of stoicheiometric ferrites. *Journal of Inorganic and Nuclear Chemistry*, **14**, 217–24.

Winkler, G. (1965). Die Bildung und Umwandlung Hexagonaler und Trigonaler Magnetischer Phasen im Dreistoffsystem $BaO-MeO-Fe_2O_3$. In *Fifth International Symposium on the Reactivity of Solids, Munich*. Ed. G. M. Schwab, Elsevier, London, pp. 572–82.

Wold, A. (1980). The preparation and characterization of materials. *Journal of Chemical Education*, **57**, 531–6.

Xu, H. K., Sorensen, C. M., Klabunde, K. J. & Hadjipanayis, G. C. (1992). Aerosol synthesis of gadolinium iron garnet particles. *Journal of Materials Research*, **7**, 712–16.

Yan, M. F. & Johnson, D. W. (1978). Impurity-induced exaggerated grain growth in Mn–Zn ferrites. *Journal of the American Ceramic Society*, **61**, 342–9.

Youshaw, R. A. & Halloran, J. W. (1982). Compaction of spray-dried powders. *American Ceramic Society Bulletin*, **61**, 227–30.

Yu, B. B. & Goldman, A. (1982). Effect of processing parameters on morphology of Mn–Zn ferrite particles produced by hydroxide–carbonate coprecipitation. In *Ferrites: Proceedings of the Third International Conference on Ferrites, 1980, Japan*. Eds. H. Watanabe, S. Iida and M. Sugimoto. Center for Academic Publications, Tokyo, pp. 68–73.

Yuan, M. S., Glass, H. L. & Adkins, L. R. (1988). Epitaxial barium hexaferrite on sapphire, by sputter deposition. *Applied Physics Letters*, **53**, 340–1.

4 Magnetic properties of ferrites

4.1 *Origin of magnetic moments*

4.1.1 *Electronic structure*

Moving electrical charges produce magnetic fields. This is a universal phenomenon, occurring for charges moving in a wire, as well as for electrons orbiting a nucleus. To understand the existence of magnetic moments in atoms, molecules and solids, analysis of atomic structure is needed. An accurate description of these structures is exceedingly complex; fortunately, examination of the simplest atom, hydrogen, provides useful insight.

Nuclear magnetic moments typically represent 10^{-3} of the electron magnetic moment and can be neglected. To a good approximation, magnetic moments in solids can therefore be described in terms of *electronic* structure alone.

A key concept in the description of electrons in atoms is their wave-like nature, introduced by de Broglie in 1925. The momentum of a particle of mass m and velocity v can be represented by a wave with wavelength λ $\lambda = h/mv$, and kinetic energy $E_c = hv$, where h is Planck's constant $(6.626 \times 10^{-34} \text{ J s})$ and v is the frequency.

Quantum mechanics is based on the wave nature of all atomic particles. In a H atom, an electron orbits around the nucleus (a proton); electron energies, or energy *states*, can be conveniently described in terms of a *wave function*, $\Psi(x, y, z, t)$, which depends on particle space coordinates, x, y, z, and time t. Stable states having well-defined (*discrete*) energies can be represented as the product of a sinusoidal time-dependent term of angular frequency ω, and a time-independent wave function $\psi(x, y, z)$:

$$\Psi(x, y, x, t) = e^{-i\omega t} \psi(x, y, z) \tag{4.1}$$

The problem is now to find which time-independent wave functions describe the energy states of the electron in a H atom. In quantum mechanics, these wave functions are found as solutions to the *Schrödinger equation*:

$$-\frac{\hbar}{2m}\left[\frac{\partial^2\psi}{\partial x^2} + \frac{\partial^2\psi}{\partial y^2} + \frac{\partial^2\psi}{\partial z^2}\right] + V\psi = i\hbar\left[\frac{\partial\psi}{\partial x} + \frac{\partial\psi}{\partial y} + \frac{\partial\psi}{\partial z}\right] \qquad (4.2)$$

where $\hbar$ is Planck's constant h divided by 2π, m the electron mass, V the electrostatic potential produced by the nucleus and, for economy, ψ represents $\psi(x, y, z)$. In spite of its apparent complexity, the Schrödinger equation expresses a simple statement (in wave-function form): the total energy of a particle, which is the right-hand side term, depends only on its kinetic energy (second-derivative term) and potential energy (*V*-term). Since the electrostatic potential produced by the nucleus has spherical symmetry, wave functions are more conveniently formulated in spherical coordinates, r, θ, and ϕ, as the product of three independent functions, R, Θ and Φ, each of them dependent on only one variable:

$$\psi(x, y, z) = R(r)\Theta(\theta)\Phi(\phi) \qquad (4.3)$$

The wave functions which are solutions to the Schrödinger equation and therefore correspond to stable energy states depend on three parameters, or *quantum numbers*:

  R depends on two parameters: n and l,
  Θ depends on two parameters: l and m_l,
  Φ depends on one parameter: m_l.

The physical meaning of these parameters is:

  n is the *principal* quantum number; it determines the total energy of the electron in each state and is given by:

$$E_n = -\left(\frac{e^2}{4\pi\varepsilon_0}\right)^2\frac{m}{\hbar^2}\frac{1}{n^2} \qquad (4.4)$$

  where e is the electron charge, ε_0 the permittivity of free space, and n a positive integer. Total energy is therefore *quantised*.

  The parameter l is related to the electron *angular momentum*, p_0, as:

$$p_0 = [l(l+1)]^{1/2}\hbar \qquad (4.5)$$

  l is also known as the *orbital* quantum number. A simplified picture of the effects of l is shown in Fig. 4.1; the eccentricity of electron orbits around the nucleus increases as l increases. Spherical orbits have $l = 0$. The *permitted* values, i.e., the restrictions on parameter l for which a solution of Schrödinger equation may be obtained, are: $l \leq n - 1$. For a given n, l can therefore have values $l = 0, 1, 2, \ldots, n - 1$.

The parameter m_l is the component of angular momentum in a given direction; it is also known as the *magnetic* quantum number because it is related to the components of the electron angular momentum when a magnetic field is applied to the atom. The permitted values are $-l \leq m_l \leq +l$.

There can exist several electron states (several combinations of quantum numbers) with the same energy (same n); these are known as *degenerate* states. A schematic representation of the electronic spatial distribution for the first three energy levels is shown in Fig. 4.2.

The experimental spectrum of atomic H shows good agreement with this model, except when it is subjected to a magnetic field, which results in a splitting of the spectrum lines. This phenomenon, also known as the *anomalous Zeeman* effect, can be explained by assuming that, in addition to its orbital momentum, an electron possesses an *intrinsic* angular momentum, p_s, with value $p_s = [s(s + 1)]^{1/2}\hbar$, where s is the *spin* quantum

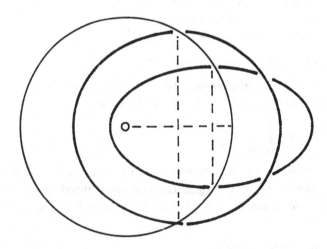

Fig. 4.1. Schematic representation of electron orbits for $n = 3$ and three different angular momenta. (Adapted from Semat & Albright, 1972.)

number. This intrinsic momentum can be visualised schematically as a spinning of the electron on its axis, with two quantised states labelled as spin *up* and spin *down*, $s = \frac{1}{2}$ and $-\frac{1}{2}$, respectively. There is no zero component possibility. This introduction of spin seems quite artificial; a rigorous derivation in which the spin appears naturally as an electron property requires relativistic wave functions.

The total angular momentum expressed in terms of the total angular quantum number, j, is $p_{tot} = [j(j + 1)]^{1/2}\hbar$, where $\mathbf{j} = \mathbf{l} + \mathbf{s}$. The magnetic (dipolar) moment, $\mathbf{M}$, of a moving particle of mass m, charge Q and angular momentum $\mathbf{A}$ is $\mathbf{M} = (Q/2m)\mathbf{A}$; atomic magnetic moments are therefore also quantised in units of $(e/2m)\hbar$. This natural unit of magnetic moment is the *Bohr magneton*, $\mu_B = 9.274 \times 10^{-24}$ A m^2. The ratio $e/2m = \gamma$ is known as the *gyromagnetic* ratio. The various magnetic moments are:

$$\mu_l = [l(l + 1)]^{1/2}\mu_B \text{ for orbital magnetic moment} \tag{4.6}$$

$$\mu_s = 2[s(s + 1)]^{1/2}\mu_B \text{ for spin magnetic moment} \tag{4.7}$$

$$\mu_j = [j(j + 1)]^{1/2}\mu_B \text{ for total angular magnetic moment} \tag{4.8}$$

The factor 2 in the spin magnetic moment appears because the values of s are half integers.

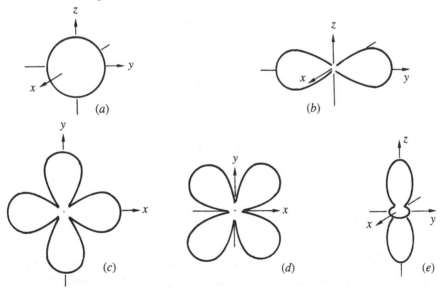

Fig. 4.2. Spatial distribution of electron density of s, p and d orbitals: (*a*) s; (*b*) p$_y$; (*c*) d$_{xy}$; (*d*) d$_{x^2-y^2}$; and (*e*) d$_{z^2}$ orbitals (schematic).

Table 4.1. *Quantum numbers for the first 60 states of H.*

n	Shell	l	m_l	s	Notation	Number of levels
1	K	0	0	$\pm\frac{1}{2}$	$1s^2$	2
2	L	0	0	$\pm\frac{1}{2}$	$2s^2$	2
		1	1	$\pm\frac{1}{2}$		
			0	$\pm\frac{1}{2}$	$2p^6$	6
			−1	$\pm\frac{1}{2}$		
3	M	0	0	$\pm\frac{1}{2}$	$3s^2$	2
		1	1	$\pm\frac{1}{2}$		
			0	$\pm\frac{1}{2}$	$3p^6$	6
			−1	$\pm\frac{1}{2}$		
		2	2	$\pm\frac{1}{2}$		
			1	$\pm\frac{1}{2}$		
			0	$\pm\frac{1}{2}$	$3d^{10}$	10
			−1	$\pm\frac{1}{2}$		
			−2	$\pm\frac{1}{2}$		
4	N	0	0	$\pm\frac{1}{2}$	$4s^2$	2
		1	1	$\pm\frac{1}{2}$		
			0	$\pm\frac{1}{2}$	$4p^6$	6
			−1	$\pm\frac{1}{2}$		
		2	2	$\pm\frac{1}{2}$		
			1	$\pm\frac{1}{2}$		
			0	$\pm\frac{1}{2}$	$4d^{10}$	10
			−1	$\pm\frac{1}{2}$		
			−2	$\pm\frac{1}{2}$		
		3	3	$\pm\frac{1}{2}$		
			2	$\pm\frac{1}{2}$		
			1	$\pm\frac{1}{2}$		
			0	$\pm\frac{1}{2}$	$4f^{14}$	14
			−1	$\pm\frac{1}{2}$		
			−2	$\pm\frac{1}{2}$		
			−3	$\pm\frac{1}{2}$		

The energy states of an electron in a H atom therefore correspond to a specific set of four quantum numbers, n, l, m_l and s. The lowest energy state (or *ground* state) is formed with the minimum values allowed for the quantum numbers, i.e., $n = 1$, $l = 0$, $m_l = 0$ and $s = 1/2$. Any excited state has $n > 1$. Each n value corresponds to a *shell*, and each l value to an *orbital*, according to spectroscopic notation. Quantum numbers for the first 60 H states are given in Table 4.1.

In atoms with more than one electron, wave functions should include the coordinates of each particle, and a new term representing the electrostatic interactions between electrons. Even for the case of only two electrons, such a wave equation is so complex that it has never been solved exactly. To analyse multielectron atoms some approximations have to be made. The most practical one is to assume that the electron considered moves in an electrical potential that is a combination of all other electrons and the nucleus, and that this potential has spherical symmetry. This approximation has proven very useful, as it allows a description of energy states in a similar manner to that employed for the H atom by using a comparable set of four quantum numbers. An important, additional condition appears: no two electrons can have the same set of quantum numbers; in other words, no more than one electron can occupy the same energy state. This is *Pauli's exclusion* principle.

The energy states of all elements in the Periodic Table can therefore be obtained by progressively 'filling' the permitted quantum numbers sets, beginning with the one-electron atom, H, followed by He, Li, etc. All complete shells (full states for a given n), as well as all full orbitals (complete occupancy of states for a given l) have zero total angular magnetic moment. This is because for any spin $s = \frac{1}{2}$ there is a $s = -\frac{1}{2}$; also, for any $l = 1$, there exists an $l = -1$. For incomplete orbitals, it is expected at first sight that only atoms with an odd atomic number (i.e., odd total number of electrons) should have one unpaired spin. However, there exist some 'irregularities' in filling electron states which affect a few elements. These irregularities in the sequence lead to large magnetic moments and to an extremely wide range of magnetic phenomena.

The electron states of elements are built up regularly until Ar ($Z = 18$) is reached with states $1s^2$, $2s^2 2p^6$, $3s^2 3p^6$, In K ($Z = 19$) the next electron, instead of occupying state $3d^1$, occupies the first state in the next shell ($n = 4$), $4s^1$ followed by $4s^2$ for Ca. The next electron then occupies the $3d^1$ level (Sc). The energy of 4s and 3d levels is therefore nearly the same, presumably because of 'screening' of the nuclear potential by other electrons. The filling sequence can be schematically written:

$$1s\ 2s\ 2p\ 3s\ 3p\ 4s\ \mathbf{3d}\ 4p\ 5s\ \mathbf{4d}\ 5p\ 6s\ \mathbf{4f}\ 5d\ 6p\ 7s\ 6d \ldots$$

Orbitals which show magnetic properties as a result of a partial filling

Magnetic properties of ferrites

Table 4.2. *Orbitals in the first transition series.*

Orbitals	K	Ca	Sc	Ti	V	Cr	Mn	Fe	Co	Ni	Cu	Zn
						Transition elements						
3d	0	0	1	2	3	5	5	6	7	8	10	10
4s	1	2	2	2	2	1	2	2	2	2	1	2

Adapted from Cullity (1972).

are highlighted. The electronic distribution of the first series of *transition* elements from Sc to Ni is shown in Table 4.2.

An additional phenomenon is that electrons show a strong tendency to occupy as many states with the *same* spin as possible. Cr and Mn, for instance, both have five electrons with spin $\frac{1}{2}$ in the 3d orbital. Electrons tend to occupy states with maximum spin angular moment, or at least with maximum total angular moment. In free atoms, the occupancy of states can be determined by *Hund's rules* (Hund, 1927; Condon & Shortley, 1935):

> *Rule 1* Electrons occupy states in each orbital with the maximum total spin number (consistent with Pauli's principle). The total spin number for the orbital, S, is calculated as $S = \sum s$. The maximum value of S corresponds therefore to an orbital which is exactly half-full.
>
> *Rule 2* The next priority is to achieve maximum total orbital momentum, L, consistent with the S value obtained by Rule 1. L is calculated as $L = \sum l$.
>
> *Rule 3* The total angular momentum, J, is calculated as $J = |L - S|$ for an orbital less than half-full, and $J = |L + S|$ for an orbital more than half-full. If the orbital is exactly half-full (five electrons in a d orbital and seven electrons in an f orbital), $L = 0$ and $J = S$.

Magnetic moments in multielectron atoms are calculated by means of the following expression:

$$\mu_{\text{eff}} = g[J(J+1)]^{1/2}\mu_{\text{B}} \tag{4.9}$$

104

Table 4.3. *Calculated and experimental magnetic moments of transition-metal ions.*

Ion	Configuration	S	L	J	$\mu_J{}^a$	$\mu_S{}^b$	Exp.
Ti^{3+}, V^{4+}	$3d^1$	1/2	2	3/2	1.55	1.73	1.8
V^{3+}	$3d^2$	1	3	2	1.63	2.83	2.8
Cr^{3+}, V^{2+}	$3d^3$	3/2	3	3/2	0.78	3.87	3.8
Cr^{2+}, Mn^{3+}	$3d^4$	2	2	0	0	4.90	4.9
Fe^{3+}, Mn^{2+}	$3d^5$	5/2	0	5/2	5.92	5.92	5.9
Fe^{2+}	$3d^6$	2	2	4	6.71	4.90	5.4
Co^{2+}	$3d^7$	3/2	3	9/2	6.63	3.87	4.8
Ni^{2+}	$3d^8$	1	3	4	5.59	2.83	3.2
Cu^{2+}	$3d^9$	1/2	2	5/2	3.55	1.73	1.9
Ce^{3+}	$4f^1$	1/2	3	5/2	2.53	1.73	2.4
Pr^{3+}	$4f^2$	1	5	4	3.58	2.83	3.6
Nd^{3+}	$4f^3$	3/2	6	9/2	3.62	3.87	3.62
Pm^{3+}	$4f^4$	2	6	4	2.68	4.90	0
Sm^{3+}	$4f^5$	5/2	5	5/2	0.85	5.92	1.54
Eu^{3+}	$4f^6$	3	3	0	0	6.93	3.6
Gd^{3+}	$4f^7$	7/2	0	7/2	7.94	7.94	8.2
Tb^{3+}	$4f^8$	3	3	6	9.72	6.93	9.6
Dy^{3+}	$4f^9$	5/2	5	15/2	10.65	5.92	10.5
Ho^{3+}	$4f^{10}$	2	6	8	10.60	4.90	10.5
Er^{3+}	$4f^{11}$	3/2	6	15/2	9.58	3.87	9.5
Tm^{3+}	$4f^{12}$	1	5	6	7.56	2.83	7.2
Yb^{3+}	$4f^{13}$	1/2	3	7/2	4.54	1.73	4.4

[a] Calculated as $g[J(J + 1)]^{1/2}$.
[b] Calculated as $2[S(S + 1)]^{1/2}$ (spin only).
Adapted from Christman (1988).

where g is the *Landé* factor, introduced to account for the difference in orbital and spin contribution to total orbital momentum:

$$g = \frac{3}{2} + \frac{S(S + 1) - L(L + 1)}{2J(J + 1)} \qquad (4.10)$$

Spin-only orbitals ($L = 0$) lead to $g = 2$. Calculated magnetic moments for free ions in the first transition series and rare earths are calculated and compared with experimental results in Table 4.3. Reasonable agreement is observed for rare-earth ions; for the first transition series, better agreement is observed with 'spin-only' estimates, as if orbital magnetic moments were cancelled. This result is discussed in the next section.

Table 4.4. *Ionisation of Fe.*

	Orbitals	Spin	Magnetic moment (μ_B) (spin only)
Atomic Fe	$4s^2 3d^6$	2	4
Fe^{2+}	$3d^6$	2	4
Fe^{3+}	$3d^5$	$\frac{5}{2}$	5
Fe^{4+}	$3d^4$	2	4

In conclusion, irregularities in the filling of electron states lead to large magnetic moments in compounds of transition elements.

4.1.2 Bonding

To form a condensed phase from isolated atoms, the electron states of the outermost orbitals have to be modified in such a way as to maximise attractive forces and minimise repulsive forces. There are four basic mechanisms for carrying out this *bonding* process, depending on the nature of the atoms involved. The four bond types are known as *ionic*, *covalent*, *metallic* and *van der Waals*. Atomic magnetic moments can be strongly modified by bonding. Since interest here is centred on ceramic materials, van der Waals bonds are not discussed.

Ionic bonding involves two types of atom; one or several electrons are typically transferred from atoms with *s* outermost orbitals, towards atoms with incomplete p orbitals. The general trend is to form ions (positive *cations* and negative *anions*) with full orbitals, therefore resulting in zero magnetic moments.

In transition-series elements, the phenomena are more complex. A first ionisation step involves transfer of the $4s^2$ electrons to form divalent cations in Sc, Ti, V, Mn, Fe, Co, Ni, and of the $4s^1$ electron in Cr and Cu to form monovalent cations. These processes should produce no change in 3d magnetic moments. Further ionisation involves 3d electrons and changes in the total spin moment are expected: it increases if the 3d orbital is more than half-full, and decreases if it is half-full or less than half-full. This can be illustrated by reference to the progressive ionisation of Fe, Table 4.4.

Origin of magnetic moments

The total magnetic moments of cations of the first transition series in solids are different to those of the corresponding atoms, even if the 3d orbitals are unchanged. The reason is that cations in solids are subjected to strong electrical fields produced by anions surrounding cation sites, i.e., they are influenced by the *crystal field* (see Sections 2.1.2 and 2.1.4). Frequently spin moments are unchanged, but orbital magnetic moments are decreased or eliminated due to the presence of the crystal field, Table 4.3. This phenomenon is referred to as *orbital quenching*; orbitals become associated with the lattice, losing their free atom character.

For certain symmetries, the crystal field can result in orbitals with decreased or even zero spin moments, independently of Hund's rules. An example is trivalent rhodium, Rh^{3+} ($Z = 45$), with $4d^6$ as the outermost occupied orbital, in an octahedral site. In a weak crystal field, Hund's rules would lead to total spin moment 4 μ_B, as shown in Fig. 4.3. The first five electrons occupy all the states of spin $\frac{1}{2}$, one in each orbital (d_{xy}, d_{yz}, d_{zx}, $d_{x^2-y^2}$ and d_{z^2}), and the sixth occupies a state $s = -\frac{1}{2}$ in d_{xy}, to give the *high-spin* configuration. If the crystal field is strong, however, occupancy

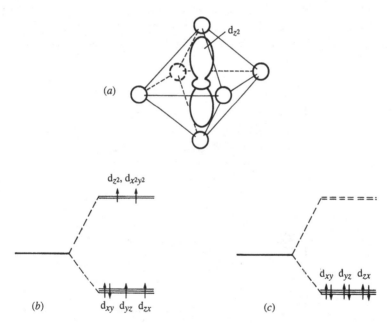

Fig. 4.3. Influence of crystal field on the electronic configuration of Rh^{3+} cations in octahedral sites: (a) geometry of the d_{z^2} orbital in an octahedral site; (b) level splitting for a weak crystal field (high spin); (c) level splitting for a strong crystal field (low spin).

of orbitals $d_{x^2-y^2}$ and d_{z^2} is prevented by the repulsion of the negative anion charge along the z-axis. The electrons are forced to occupy the available states, with spin $-\frac{1}{2}$, in orbitals d_{yz} and d_{zx}. The Pauli principle leads to antiparallel spins in the three occupied orbitals and therefore to zero spin moment. This is the *low spin* configuration. These effects are virtually absent in rare-earth cations because orbitals with magnetic moments (4f) are shielded from the crystal field by 5s and 5p orbitals; total magnetic moments show a good agreement with those calculated from atomic levels, Table 4.3.

Covalent bonding is characterised by 'sharing' electrons between atoms; each covalent bond is formed by two electrons, one from each atom, with opposite spins. It is a directional bonding, depending strongly on the geometry of the orbitals involved. Covalent bonding is generally considered to have little effect on the magnetic moments of transition-series atoms; however, purely covalent or ionic bonds are rare and intermediate bonds (incomplete transfer of electrons) although very common, are difficult to characterise.

Metallic bonding can be briefly described by considering the interactions between atoms of the same type, when they are in close proximity, Fig. 4.4. As a result of interactions, the original atomic levels *split* into new levels, with a small difference in energy between them at the

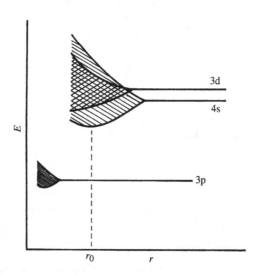

Fig. 4.4. Atomic level splitting into energy bands to form a metallic bond. The equilibrium interatomic distance is r_0.

equilibrium interatomic separation. Interaction of N atoms leads to $2N$ band levels (N with spin up and N with spin down) for an s atomic orbital, $6N$ band levels for a p orbital, $10N$ band levels for a d atomic orbital, etc. This phenomenon is expected to occur for any atomic orbital; core levels, such as the 3p level in Fig. 4.4, would also split into a band if interatomic separations were reduced further.

From a quantum-mechanical point of view, it is said that electron wave functions are 'spread out' over the entire solid; atomic levels are transformed into an *energy band*, where differences between energy levels are so small that it can be thought of as a continuous distribution of states. Electrons in a band can no longer be assigned to a particular cation. They are *delocalised* and become free, in the sense that having energy states available, they can easily be excited by an external electrical field to transport current, for example.

The occupancy of atomic levels determines the occupancy or filling of a band, which, in turn, is crucial regarding delocalisation of the electrons. N atoms with outermost s^1 levels (thus N s electrons) lead to s bands with $2N$ states, which are therefore half-filled. The resulting solid is a good electrical conductor. The energy of the highest occupied level is known as the *Fermi* energy, or the Fermi level, E_f. Since N atoms form a solid with an s band with $2N$ states available, N atoms with two electrons in an s level result in an insulator. This is because all the states in the band are occupied; the Fermi level corresponds to the band limit, as illustrated in Fig. 4.5. In most cases, there is an energy difference between the filled band and the next, empty band. The occupied band is known as the *valence* band, the empty one is the *conduction* band and the energy separation between them is the energy *gap* or the *forbidden* band. The last name implies that there are no available states for electrons in this energy range. A general classification of solids into conductors, insulators and semiconductors is based on the value of the energy gap.

There are some cases, however, in which s^2 atoms form conductor solids. In these metals, such as Mg for instance, the level splitting occurs in such a way that the Fermi level (the highest energy level in the s band) overlaps with the lowest level of the p band. This means that, although the s band is full, electrons can easily be excited into the available states in the p band. The energy gap is zero. Band theory is a general theoretical approach for solid-state analysis; ionic and covalent bonding can also be described in terms of energy bands.

Again, the case of transition metals is more complex. As a result of the

Table 4.5. *Experimental values of magnetic moments in transition metals.*

Metal	Atomic levels	Calculated spin only (μ_B)	Experimental (μ_B)
Fe	$3d^6 4s^2$	4	2.2
Co	$3d^7 4s^2$	3	1.72
Ni	$3d^8 4s^2$	2	0.60

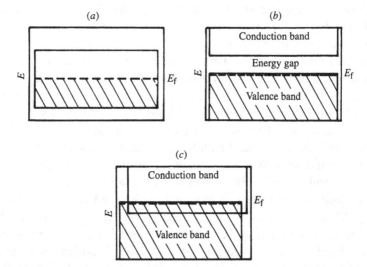

Fig. 4.5. Energy bands: (*a*) Fermi level in the middle of the conduction band, leading to a conductor; (*b*) full valence band, with Fermi energy at the top, resulting in an insulator or semiconductor; and (*c*) full valence band but overlapping with the conduction band, leading to a conductor.

similarity in the energies of the 3d and 4s atomic levels, splitting leads to overlap of the 3d and 4s bands. Since transition metals have partially filled 3d orbitals, they are good conductors. Regarding their magnetic behaviour, it is expected that spin-up atomic levels lead to spin-up bands, and conversely for spin-down levels, resulting in an intrinsic magnetic moment per atom. Experimental measurements in Fe, Co and Ni, Table 4.5, confirm the existence of a magnetic moment per atom, but surprisingly, all of them exhibit non-integral values in units of Bohr

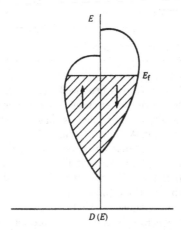

Fig. 4.6. Schematic density of states for the two spin states in a solid produced by atoms with unequally filled levels, leading to a non-integral number of Bohr magnetons per band.

magnetons. A concept which is very useful when discussing band problems is the *density of states*, which is a representation of the number of states a band holds in each energy interval, Fig. 4.6. Bands formed by unequally filled levels show a difference in filling in spin-up and spin-down bands; the Fermi level is the same in both parts of the band. However, the energy difference between spin-up and spin-down bands, ΔE, is determined not only by the difference in filling of atomic levels but by a variety of factors, such as crystal symmetry, lattice parameters, etc. Electrons begin to fill the lowest energy levels in the spin-up band and occupy levels in the spin-down band before the other half is full, Fig. 4.6. Since levels are closely spaced (in a ~ 1 g sample of metal, a band has $\sim 10^{23}$ energy levels), this leads to a non-integral value of net magnetic moment. These results illustrate the essential differences between atomic (or ionic) levels and band levels.

4.2 Magnetic order

4.2.1 Diamagnetism and paramagnetism

A useful property for characterising magnetic materials is the *magnetic susceptibility*, χ, defined as the *magnetisation*, M, divided by the applied

Table 4.6. *Magnetic units in the SI system.*

Parameter	Definition and units
H	Magnetic field strength. Unit: ampere per metre, A/m. A field of 1 A/m is produced by an infinitely long solenoid with n turns per metre of coil, carrying a current of $1/n$, A.
M	Magnetisation; defined as magnetic moment per volume unit. Unit: A/m. (SI-Sommerfeld.)
I	Intensity of magnetisation; same definition, but $I = \mu_0 M$, where μ_0 is the permeability of free space, $4\pi \times 10^{-7}$ H/m; I therefore has the unit of the Tesla, T, or weber/m^2 (Wb/m^2). (SI-Kennelly.)
B	Magnetic induction. Unit: T (Wb/m^2). A magnetic induction 1 T generates a force of 1 N/m on a conductor carrying a current of 1 A, perpendicular to the direction of induction.
Basic equation:	$B = (H + M)$ (Sommerfield)
	$B = \mu_0 H + I$ (Kennelly)
Φ	Magnetic flux: $\Phi = BA$, where A = cross-sectional area. Unit: Wb.
χ	Magnetic susceptibility; $\chi = M/H$. Dimensionless since both M and H are given in A/m.
μ	Magnetic permeability, B/H. Unit: H/m (Henry/m).
μ_r	Relative permeability, μ/μ_0. Dimensionless.
	$\mu_r = 1 + \chi$.

magnetic field, H:

$$\chi = M/H \qquad (4.11)$$

Definitions, units and conversion factors of important magnetic parameters are given in Tables 4.6, 4.7 and 4.8. A comprehensive discussion of magnetic units has been given by Brown (1984). As magnetisation is a volume property, χ is also known as the volume susceptibility. It can be expressed as mass and molar susceptibility with appropriate conversion factors, Table 4.7. The three parameters in Eq. (4.11) are vectors; however, since they tend to be parallel, this equation is usually written in scalar form. Magnetic moments tend to align in the same direction as an applied field because a parallel configuration leads to decreased magnetic

Table 4.7. *Magnetic units in the cgs system.*

H	Magnetic field strength; oersted (Oe). A field of 1 Oe exerts a force of 1 dyne on a unit pole. (The cgs system is based on magnetic poles.)
M	Magnetisation; magnetic moment per unit volume; erg/Oe cm^3, or emu/cm^3 ('emu' = electromagnetic unit = erg/Oe). When magnetisation is expressed as $4\pi M$, the unit is the gauss (G).
σ	Specific magnetisation, M/d, with d = density in g/cm^3. The unit of σ is therefore emu/g.
B	Magnetic induction; gauss (G).
Basic equation:	$B = H + 4\pi M$
Φ	Magnetic flux, BA, with A = cross-sectional area in cm^2. The unit of Φ is the maxwell.
χ	Magnetic susceptibility, M/H. Unit: emu/Oe cm^3.
χ_m	Mass susceptibility, χ/d. Unit: emu/Oe g.
χ_M	Molar susceptibility, $\chi_m W$, where W is the molecular weight. Unit: emu/Oe g mol.
μ	Magnetic permeability, B/H. Dimensionless; in addition, as $\mu_0 = 1$ in the cgs system, μ has the same value as the relative permeability, μ_r, in the SI system. In the cgs system, the permeability and the susceptibility are related by: $\mu = 1 + 4\pi\chi$

energy:

$$E_p = -\boldsymbol{\mu} \cdot \boldsymbol{H} = -|\mu||H| \cos\theta \qquad (4.12)$$

where μ is a magnetic moment and θ the angle between it and applied field. E_p is also known as the magnetic potential energy.

The temperature dependence of susceptibility or, more accurately, the inverse of susceptibility is a good characterisation parameter. A small, negative, constant value, Fig. 4.7, corresponds to a *diamagnetic* material; a linear relationship is shown by a *paramagnetic* material. In some metals, known as *Pauli* paramagnets, a small, positive, constant susceptibility is observed. Some other materials show linear behaviour only for high temperatures; below a critical temperature, their susceptibility shows a complex behaviour. They are called ferro-, ferri- or antiferromagnets.

Diamagnetic materials have no intrinsic magnetic moments. Negative

Table 4.8. *Conversion factors between systems.*

Field strength, H	1 Oe = 79.577 A/m
Magnetisation, M	1 emu/cm^3 = 1000 A/m
Intensity of magnetisation, I	1 emu/cm^3 = 12.57 × 10^{-4} T
Magnetisation $4\pi M$ (cgs) → I(SI)	1 G = 10^{-4} T
Induction, B	1 G = 10^{-4} T
Magnetic flux, Φ	1 maxwell = 10^{-8} Wb
Permeability μ(cgs) → μ_r(SI)	$\mu = \mu_r$ (both dimensionless)
Energy product, $(BH)_{max}$	MGOe = 7.958 kJ/m^3
Anisotropy constant, K_1	erg/cm^3 = 10^{-1} J/m^3
Domain wall energy, γ_w	erg/cm^2 = 10^{-3} J/m^2

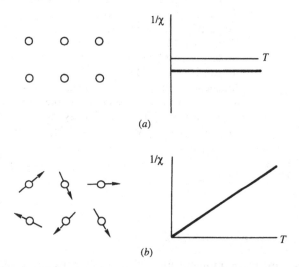

Fig. 4.7. Characteristic temperature dependence of the reciprocal susceptibility: (*a*) for a diamagnetic and (*b*) for a paramagnetic material. (Adapted from Cullity, 1972.)

susceptibility results from Lenz's law: the change produced by an external field on electronic currents induces an emf which opposes the field. Organic compounds are usually diamagnetic. Typical susceptibility values for diamagnetic materials are in the -1×10^{-7} to -2×10^{-6} range. Superconductors are perfect diamagnets, with $\chi = -1$.

Paramagnets possess intrinsic magnetic moments; an applied field tends to align them, but thermal agitation promotes randomisation. The classical theory of paramagnetism was developed by Langevin in 1905, based on the reduction in magnetic energy (Eq. (4.12)), and a probability with Boltzmann form $e^{-E_P/kT}$, where k is Boltzmann's constant. The derivation leads to:

$$M = N\mu \coth a - 1/a \qquad (4.13)$$

where N is the number of atoms per unit volume, μ is the atomic magnetic moment and a is $\mu H/kT$. The right-hand side of Eq. (4.13) is known as the *Langevin* function, and can be expressed as a series:

$$L(a) = \coth a - 1/a = a/3 - a^3/45 + 2a^5/945 - \cdots \qquad (4.14)$$

For high a values, i.e., a large applied field and/or a low temperature, $L(a) \approx 1$ and $M_0 \approx N\mu$; this is known as *saturation*, because $N\mu$ corresponds to the maximum magnetisation state, when all magnetic moments are aligned parallel to the field. However, saturation is never achieved in paramagnetic materials; the susceptibility is so small that even with the highest fields available, $M/N\mu$ is only 1 or 2%.

At the other extreme, when a is small, $L(a) \approx a/3$ and Eq. (4.14) becomes:

$$M = N\mu\mu H/3kT \qquad (4.15)$$

from which Curie's law follows:

$$\chi = N\mu^2/3kT = C/T \qquad (4.16)$$

where C, the Curie constant, is $N\mu^2/3k$. Figure 4.7(b) is a plot of Eq. (4.16).

Application of quantum mechanics to paramagnetism results in a better quantitative agreement with experiment, but the form of Eq. (4.16) is not changed. In this theory, instead of a system with a continuous distribution of available energy states, a discrete-level system is considered. The general expression for magnetisation is:

$$M/M_o = \frac{2J + 1}{2J} \coth\left(\frac{2J + 1}{2J}\right)a' - \frac{1}{2J} \coth\frac{a'}{2J} \qquad (4.17)$$

where $a' = gJ\mu_B H/kT$. The right-hand side is the *Brillouin* function. When $J = \infty$, the classical (Langevin) expression is obtained; for $J = \frac{1}{2}$, the spin-only case, expression (4.17) becomes:

$$M/M_0 = \tanh a' \qquad (4.18)$$

For small values of a', this expression reduces to:

$$\chi = N\mu_{eff}^2/3kT \tag{4.19}$$

with $\mu_{eff} = g[J(J + 1)]^{1/2}\mu_B$ and $C = N\mu_{eff}^2/3k$.

Pauli paramagnets are metals with a small, positive, constant susceptibility. This is a consequence of the band structure of energy levels in metals; not all electrons in a band are allowed to invert their spin when an external field is applied. Only electrons near the limit of the band (near the Fermi energy level) are able to occupy states with inverted spin. This fraction can be estimated roughly as $\sim kT/E_f$; if the Curie law is still applicable, it has to be multiplied by this fraction:

$$\chi = (C/T)(kT/E_f) = kC/E_f = \text{constant} \tag{4.20}$$

since E_f is constant.

4.2.2 Exchange

In addition to showing large, intrinsic magnetic moments, many materials containing transition elements behave as if they were spontaneously magnetised. Fe, Co, Ni, Gd and many of their alloys, as well as many of their oxides and fluorides exhibit spontaneous magnetic ordering. Various types of magnetic moment ordering have been observed, Fig. 4.8: parallel or *ferromagnetic*; antiparallel, which can lead to *antiferromagnetism* if the moments are exactly compensated, or *ferrimagnetism* if moments are antiparallel but uncompensated, giving rise to a resultant moment; other more complex ordering phenomena such as *canted* (or triangular) and *helical* have also been observed.

The interactions responsible for these phenomena are strong, since the critical temperatures at which the magnetic order is destroyed by thermal agitation can be as high as 1400 K (Co). An interaction based on purely magnetic forces, such as the effect of the magnetic field produced by one spin on its nearest neighbours is extremely small; it would give rise to maximum critical temperatures around 1 K. The physical origin of these interactions is electrostatic, quantum-mechanical, and magnetic ordering is only a natural consequence.

To have some insight into the origin of magnetic ordering, it is instructive briefly to analyse the most simple system: two atoms a and b, with one electron each, close enough to give rise to electron interaction. Electron wave functions for the interacting system can be expressed as linear combinations of the original atomic wave functions; this is known as the Heitler–London approximation. The total energy can then be expressed as:

$$E = E_a + E_b + Q \pm J_{ex} \tag{4.21}$$

where E_a and E_b are the energies of electrons when they orbit their separate atoms a and b, Q is the electrostatic (coulomb) interaction energy, and a new term appears, J_{ex}, referred to as the *exchange* energy, or the exchange integral. J_{ex} arises from the possibility of exchange between electrons, when electron a moves around nucleus b and electron b orbits nucleus a. They are indistinguishable, except for their spins. Their relative orientation is therefore the most important factor; parallel spins lead to positive J_{ex}, and antiparallel to negative J_{ex}. Exchange energy is a consequence of the Pauli principle, when it is applied to an atomic system.

By applying the Heitler–London approximation to transition metals, Bethe (1933) calculated exchange integrals for Fe, Co, Ni, Cr and Mn, as a function of interatomic distance and radii of 3d orbitals,

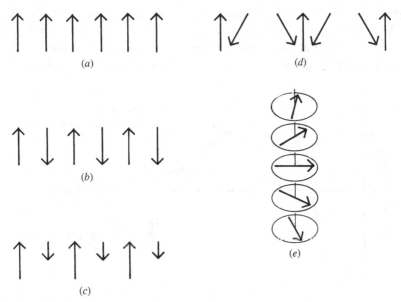

Fig. 4.8. Schematic representation of some magnetic structures: (*a*) ferromagnetic; (*b*) antiferromagnetic, (*c*) ferrimagnetic; (*d*) triangular or canted; and (*e*) helical.

117

Fig. 4.9. This plot, known as the Bethe–Slater curve, gives the correct sign for J_{ex}; Fe, Ni and Co are ferromagnetic (parallel, $J_{ex} > 0$); Mn and Cr are antiferromagnetic (antiparallel, $J_{ex} < 0$). The shortcomings of this model have been discussed by Herring (1962); however, it remains as an instructive approximation.

In 1928, Heisenberg showed that, for many practical purposes, exchange energy can be represented by the relation:

$$E_{ex} = -2J_{ex}\mathbf{s}_1 \cdot \mathbf{s}_2 = -2J_{ex}s_1s_2 \cos \theta \qquad (4.22)$$

where E_{ex} is the exchange energy dependent on the relative orientation of one pair of nearest neighbour spins, s_1 and s_2, and θ is the relative angle between them. For $J_{ex} > 0$, ferromagnetic order results in an energy minimum; for $J_{ex} < 0$, an antiparallel spin arrangement is favoured. For a solid, a summation over all spin pairs is necessary to calculate the total exchange energy.

The Curie temperature, T_C, and the spontaneous magnetisation, or *saturation magnetisation*, M_s, for some ferromagnetic materials are given in Table 4.9. Magnetic moments do not disappear at the transition temperature, but become disordered. On increasing temperature, ferromagnetic materials undergo a transition from an ordered ferromagnetic phase for $T < T_C$, to a disordered, paramagnetic phase at $T > T_C$. Thermodynamically, this is a second-order transition. The additional enthalpy contribution necessary to disorder the magnetic moment alignment is provided over the whole temperature range from 0 K to T_C,

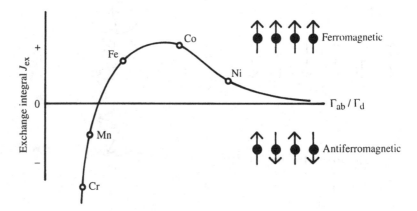

Fig. 4.9. The Bethe–Slater curve, defining the sign of the exchange integral, J_{ex}, for transition metals. (Adapted from Cullity, 1972.)

Table 4.9. *Curie temperature and saturation magnetisation of transition metals.*

Metal	T_C (K)	$M_s(0)$ (kA/m)	M_s (293 K) (kA/m)
Fe	1043	1740	1710
Co	1404	1430	1420
Ni	631	510	480

Adapted from Cullity (1972).

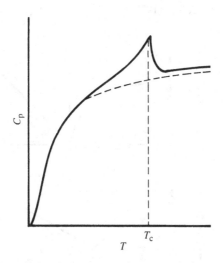

Fig. 4.10. Schematic plot of the magnetic contribution to the heat capacity in a magnetic solid, as a function of temperature.

Fig. 4.10, instead of in the form of latent heat at the transition. This results in a magnetisation curve with small variations at low temperatures and a large decrease as T approaches T_C, Fig. 4.11. The concepts of *reduced* magnetisation and *reduced* temperature, $m = M(T)/M(0)$, and $t = T/T_C$, respectively, are used to compare materials with different spontaneous magnetisations and Curie points. $M(T)$ represents the magnetisation value at temperature T, and $M(0)$ the value at 0 K. Since in ferromagnets magnetisation is always maximum for 0 K, a decreasing curve is obtained.

Table 4.10. *Characteristic temperatures of some antiferromagnetic oxides.*

Material	T_N (K)	θ (K)
α-Fe_2O_3	950	2000
NiO	523	3000
CoO	293	280
FeO	198	570
MnO	122	610

Adapted from Cullity (1972).

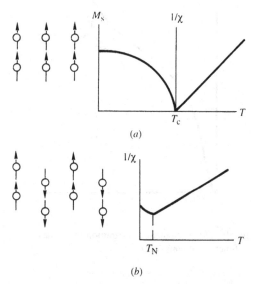

(a)

(b)

Fig. 4.11. Temperature dependence of the inverse susceptibility for: (a) a ferromagnetic solid, showing a spontaneous magnetisation for $T < T_C$ and Curie–Weiss behaviour for $T > T_C$; (b) an antiferromagnetic material; T_N is the Néel temperature. (Adapted from Cullity, 1972.)

Magnetic ordering mechanisms in oxides result mostly in antiparallel spin arrangements and are considerably more complex than in metals. MnO, FeO, and CoO are all *antiferromagnets* at temperatures below room temperature; NiO has a transition temperature of 523 K and α-Fe_2O_3, 950 K, Table 4.10. Transition temperatures in antiferromagnets are known as 'Néel temperatures', because L. Néel (1932) was the first to develop a theory to explain their properties.

The problem in oxides is that an 'indirect' exchange has to be assumed, which gives rise to ordering of cations separated by an O^{2-} (or an anion in general). Cations are too far apart in most oxides for a direct cation–cation interaction. An essentially qualitative model for this indirect interaction, also referred to as *superexchange*, is as follows. Consider two transition metal cations separated by an O, Fig. 4.12. The O^{2-} has no net magnetic moment since it has completely filled shells, with p-type outermost orbitals. Orbital p_x has two electrons: one with spin up, and the other with spin down, consistent with Pauli's principle. When one of the transition-metal cations is brought close to the O^{2-}, partial electron overlap (between a 3d electron from the cation and a 2d electron from the O^{2-}) can occur only for antiparallel spins, because electrons with the same spin are repelled. Empty 3d states in the cation are available for partial occupation by the O^{2-} electron, with an antiparallel orientation. Electron overlap between the other cation and the O^{2-} then occurs resulting in antiparallel spins and therefore antiparallel order between the cations. Since p orbitals are linear, the strongest interaction is expected to take place for cation–O^{2-}–cation angles close to 180°, as has been widely observed. A general, semiempirical approach has been discussed by Anderson (1963).

4.2.3 Molecular field theory

Molecular field theory (MFT) is the simplest theory of ordered magnetic moments in solids. Proposed in 1907 by P. Weiss, its main assumption is

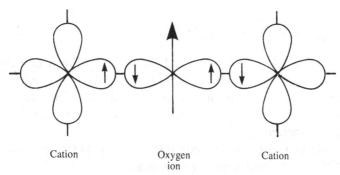

Cation   Oxygen   Cation
ion

Fig. 4.12. Schematic representation of the superexchange interaction in magnetic oxides. The p orbitals of an anion (centre) interact with the d orbitals of transition-metal cations.

that ordering results from a fictive field, the *molecular field*, which depends on magnetisation:

$$H_w = \lambda M \qquad (4.23)$$

where H_w is the molecular field, M the magnetisation and λ the molecular field coefficient. The total field experienced by magnetic moments is then:

$$H_T = H_w + H \qquad (4.24)$$

where H is the external field. The Curie law is now:

$$\chi = C/T = M/H_T \qquad (4.25)$$

Replacing H_w by λM and solving for M:

$$M = CH/(T - \lambda C) \qquad (4.26)$$

with $\lambda C = T_C$. Expressing the susceptibility as a function of only the external field leads to:

$$\chi = C/(T - T_C) \qquad (4.27)$$

which is the Curie–Weiss law. This leads to a linear behaviour of $1/\chi$ for $T > T_C$, as observed in many ferromagnetic materials. This expression has another implication: at $T = T_C$, χ becomes infinity; since $\chi = M/H$, it can be interpreted as the existence of finite magnetisation at zero field, i.e., a spontaneous magnetisation.

A simple estimate shows that magnetic order is not the result of a real magnetic field. The order of magnitude of the magnetic interaction energy for one spin, $\mu_B H_w$, can be compared with the thermal energy at the Curie transition, kT_C:

$$\mu_B H_w \approx kT_C; \qquad H_w \approx kT_C/\mu_B \qquad (4.28)$$

Taking $\mu_B = 1.165 \times 10^{-29}$ J m/A, $k = 1.381 \times 10^{-23}$ J/K and $T_C \approx 1000$ K (T_C for Fe is 1040 K), leads to $H_w \approx 1.2 \times 10^9$ A/m. The actual field produced by a spin on its neighbours is $\approx 8 \times 10^4$ A/m. The Weiss molecular field is not a real field, but a simplified representation of the exchange interaction.

The molecular field can be introduced in Eq. (4.18) to investigate thermal effects on spontaneous magnetisation:

$$M/M_0 = \tanh a = \tanh(\mu_B H_w/kT) \qquad (4.29)$$

where the external field is zero ($H_T = H_w$) and the spin-only contribution is taken ($g = 2$ and $J = \frac{1}{2}$). Since $H_w = \lambda M$, the hyperbolic tangent argument becomes $a' = \mu_B \lambda M/kT$. If $T_C = N\mu_B^2\lambda/k$, $a' = (M/M_0)/(T/T_C) = m/t$. Eq. (4.29) can be written:

$$m = \tanh(m/t) \tag{4.30}$$

which can be solved numerically. It is plotted in Fig. 4.13, and compared with reduced magnetisation curves of Fe, Co, Ni, as well as $NiFe_2O_4$. A good agreement is observed for ferromagnets; the ferrite curve deviates considerably, however. Ferrite behaviour is more complex since spontaneous magnetisation in ferrites corresponds to the difference between magnetisations of two or more magnetic *sublattices*. All the sublattices have the same Curie temperature, but each sublattice can have its own temperature dependence.

Antiferromagnetic materials have two identical magnetic sublattices. Their resultant susceptibility is small and shows a maximum at the Néel temperature, Fig. 4.14. MFT is applied to antiferromagnets by considering

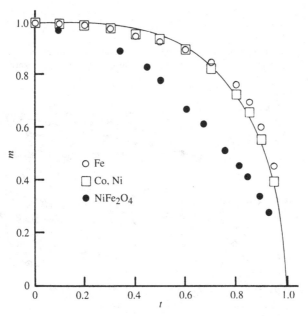

Fig. 4.13. Magnetisation as a function of temperature for metallic Fe, Ni, Co and $NiFe_2O_4$, in terms of reduced parameters: $m = M(T)/M(0)$ and $t = T/T_C$. (Ferrite data from Pauthenet, 1952.)

two molecular fields acting on each sublattice:

$$H_A = \lambda_{AA} M_A - \lambda_{AB} M_B \Big\}$$
$$H_B = \lambda_{BB} M_B - \lambda_{BA} M_A \Big\}$$

(4.31)

The minus sign accounts for the antiparallel order between sublattices (λ is taken as positive), Fig. 4.15. Susceptibility can be expressed by:

$$\chi = C/(T + \theta)$$

(4.32)

In a $1/\chi$ plot, Fig. 4.14, θ is the extrapolation of the linear relation for

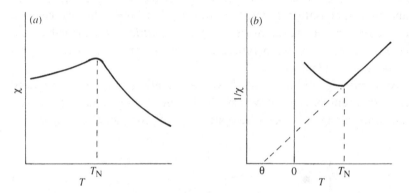

Fig. 4.14. (*a*) Susceptibility of an antiferromagnetic material as a function of temperature and (*b*) its inverse. T_N and θ are the Néel and asymptotic temperatures, respectively (schematic).

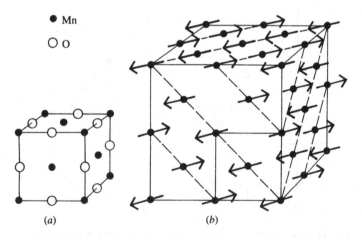

Fig. 4.15. Antiferromagnetic order in MnO: (*a*) crystallographic unit cell; and (*b*) magnetic unit cell. (Adapted from Cullity, 1972.)

$T > T_C$ on the T-axis. Values of θ for some antiferromagnets are given in Table 4.10. The paramagnetic susceptibility of antiferromagnets follows a Curie–Weiss law with negative critical temperature. At $T < T_C$, susceptibility depends on the orientation of the applied field relative to the antiparallel axis; after some approximations, when the external field is perpendicular to this axis, it can be shown that:

$$\chi_\perp = C/2\theta \tag{4.33}$$

whereas for a parallel applied field:

$$\chi_\| = \frac{2N\mu^2 B'(J, a')}{2kT + 2N\mu^2 B'(J, a')} \tag{4.34}$$

where N is the number of magnetic moments per unit volume, μ their magnitude, B' is the derivative of Brillouin function (Eq. (4.17)) with respect to its argument a'; $a' = \mu H/kT$. Perpendicular, $\chi_\perp$, and parallel, $\chi_\|$, susceptibilities for a MnF$_2$ single crystal are plotted in Fig. 4.16, and compared with results on a polycrystalline sample, χ_p. A detailed discussion of the application of MFT to antiferromagnets is given by Smart (1966). MFT has had considerable success in investigations of antiferromagnetic solids since this theory assumes localised magnetic moments, and most antiferromagnets are insulators or semiconductors.

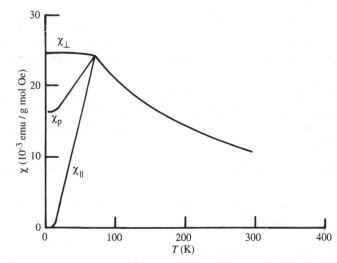

Fig. 4.16. Molar susceptibility of MnF$_2$ as a function of temperature. $\chi_\perp$ and $\chi_\|$ were measured in perpendicular and parallel fields, respectively, on a single crystal. χ_p refers to a polycrystalline sample. (Bizette & Tsai, 1954.)

MFT has also been applied to ferrites. The complexity of the mathematical treatment increases rapidly since, in spinels and garnets, in the simplest case, two different, antiparallel sublattices have to be considered; in hexagonal ferrites, there are at least four sublattices.

In the most simple case, with two sublattices in a spinel, molecular fields are expressed in a similar way to that used for antiferromagnetics; the difference is that now there is a net magnetic moment, because the sublattice magnetisations do not cancel each other. The net magnetisation at temperature T (at $T < T_C$) is simply:

$$M(T) = M_{oct}(T) - M_{tet}(T) \tag{4.35}$$

where $M_{oct}(T)$ and $M_{tet}(T)$ are the magnetisation of octahedral and tetrahedral sublattices, respectively, at temperature T. For $T < T_C$, the Curie–Weiss expression becomes (see Cullity (1972) for instance):

$$\frac{1}{\chi} = \frac{T}{C} + \frac{1}{\chi_0} - \frac{b}{T - \theta} \tag{4.36}$$

where χ_0, θ and b are related to molecular field coefficients. At high temperatures, the last term on the right-hand side becomes negligible and:

$$\frac{1}{\chi} = \frac{T}{C} + \frac{1}{\chi_0} \tag{4.37}$$

which is shown in Fig. 4.17. Experimental results for Mg ferrite, Fig. 4.18, are in good agreement, except near the Curie point. This difference is generally attributed to short-range order of magnetic moments persisting just above T_C.

Rigorous application of MFT can be made to general cases with defined sublattices (long-range ordered structures) (Fuentes, Aburto & Valenzuela, 1987). Net magnetisation is given by the resultant of contributions from all the sublattices:

$$M_T(T) = \sum_i M_i(T) \tag{4.38}$$

The magnetisation for each sublattice is obtained from generalised molecular field expressions:

$$M_i(T) = M_i(0)B_i\left[(1/kT) \sum_i Z_{ij}\lambda_{ij}s_iM_j \right] \tag{4.39}$$

where $M_i(0)$ is the magnetisation of sublattice i at 0 K, B_i the Brillouin

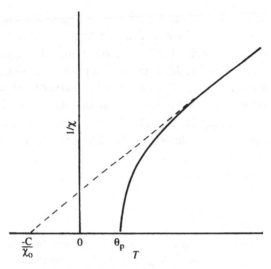

Fig. 4.17. The temperature dependence of the inverse susceptibility for ferrimagnets, according to the MFT.

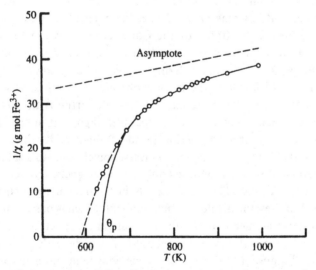

Fig. 4.18. The inverse susceptibility of Mg ferrite as a function of temperature for $T > T_C$, referred to one mole of Fe^{3+}. The solid line and θ_p correspond to Eq. (4.33). (Adapted from Cullity, 1972; data from Serres, 1932; curve constants from Néel, 1948.)

function for spin i, Z_{ij} the number of nearest neighbours j of sublattice i, λ_{ij} the molecular field coefficient which transforms the magnetisation from sublattice j into a molecular field acting on sublattice i, s_i the spin value in sublattice i, and finally, M_i is the magnetisation of sublattice i. This approach has been applied to spinels (two-sublattices: Stadnik & Zarek (1979); Srivastava *et al.* (1976); three-sublattices, Aburto *et al.* (1982)) and garnets (two-sublattices: Dionne (1970); Röschmann & Hansen (1981)). The shortcoming of this theory is that for three or more sublattices, mathematical fitting can generate virtually any curve, regardless of its physical meaning.

4.2.4   Ferrimagnets

Spontaneous magnetisation of *spinels* (at 0 K) can be estimated on the basis of their composition, cation distribution and the relative strength of the possible interactions. Since cation–cation distances are generally large, direct (ferromagnetic) interactions are negligible. Because of the geometry of orbitals involved, the strongest superexchange interaction is expected to occur between octahedral and tetrahedral cations, or A–O–B interactions, with an angle $\sim 135°$, Fig. 2.2. The next interaction is B–O–B, where cations make a $90°$ angle. The relative strengths of these two interactions are illustrated by the Curie temperatures of Li and Zn ferrites; the former has a cation distribution $\overrightarrow{Fe}[Li_{0.5}\overleftarrow{Fe}_{1.5}]O_4$, and the latter $Zn[\overrightarrow{Fe}\overleftarrow{Fe}]O_4$, where spin ordering is indicated by arrows. The Curie temperature for Li ferrite is the highest observed in spinels, 958 K; the Néel temperature (it is antiferromagnetic) for Zn ferrite is the lowest of the ferrites, 9 K. Direct evidence of antiparallel alignment in spinels was first obtained by neutron diffraction (Shull, Wollan & Koehler, 1951).

  The A–O–B dominant interaction in inverse spinels leads to a saturation magnetisation at 0 K which depends only on the magnetic moment of the divalent cation. Antiparallel order cancels Fe^{3+} moments in octahedral and tetrahedral sites; the divalent cation moment, which is parallel to Fe in octahedral sites then accounts for the net resultant magnetisation. In the case of Mn ferrite, saturation magnetisation is independent of cation distribution, because Mn^{2+} has the same magnetic moment per ion $(\sim 5\mu_B)$ as Fe^{3+}. Calculated and experimental values of saturation magnetisation per formula unit for some ferrites are given in Table 4.11. To transform the magnetic moment (in Bohr magnetons/formula unit) to macroscopic magnetisation units (A/m), it is important to remember that

Table 4.11. *Magnetisation and Curie temperature of some spinels.*

	T_C (K)	$n(\mu_B)$ Calc.	$n(\mu_B)$ Exp.	M_s (0 K) (kA/m)	M_s (300 K) (kA/m)
$Fe^{3+}[Fe^{2+}Fe^{3+}]$	858	4	4.1	510	480
Fe[NiFe]	858	2	2.3	300	270
Fe[CoFe]	793	3	3.7	475	425
$MnFe_2{}^a$	573	5	4.6	560	400
Fe[CuFe]	728	1	1.3	160	135
$Zn[Fe_2]$	9	0	0	0	0

a Magnetic moment independent of cation distribution.
Adapted from Smit & Wijn (1961).

magnetisation is defined as the magnetic moment per unit volume:

$$M_s = 8n\mu_B/V \qquad (4.40)$$

where n is the number of magnetic moments per formula unit MFe_2O_4 and V is the unit cell volume. For magnetite, for instance, the calculated $n = 4$ (Fe^{2+} is $3d^6$), the unit cell volume is $(8.394 \times 10^{-10} \text{ m})^3$ and $\mu_B = 9.274 \times 10^{-24}$ A m^2, which leads to $M_s = 5.02 \times 10^5$ A/m. The measured magnetisation is 5.1×10^5 A/m, which is ~ 4.1 Bohr magnetons per formula unit. The difference ($0.1\mu_B$) is usually attributed to unquenched orbital momentum.

The temperature dependence of magnetisation in spinels is, in fact, the difference in sublattice magnetisations. Since each sublattice has its own temperature dependence, a wide variety of magnetisation curves can be obtained in ferrites. Temperature effects can involve a change in configuration, say from antiparallel to canted or 'triangular' in a particular temperature range, but there is only one transition temperature from an ordered to the paramagnetic phase. Spontaneous magnetisation is an essentially collective phenomenon; any particular spin is ordered (oriented in a particular direction with respect to all others) because a 'molecular' field, produced by all the other spins influences it. Magnetisation curves for some spinels are shown in Fig. 4.19.

Magnetic properties can be modified widely by cation substitution. An illustrative case is substitution of Ni by Zn in Ni ferrite to form solid solutions $Zn_{1-x}Ni_xFe_2O_4$. Ni ferrite is an inverse spinel, with Fe in A sites; Ni and the remaining Fe share B sites. Zn^{2+} preferentially enters the A sites (see Chapter 1). The cation distribution can be written $Zn_xFe_{1-x}[Ni_{1-x}Fe_{1+x}]O_4$. Zn^{2+} is diamagnetic ($3d^{10}$) and its main effect is to break linkages between magnetic cations. Another effect is to increase intercation distances by expanding the unit cell, since it has an ionic radius larger than the Ni and Fe radii. The Curie temperature decreases steeply as a function of Zn content, Fig. 4.20. By comparing Ni–Zn and Ni–Cd solid solutions, it is possible to evaluate the effect of the intercation distance on the Curie temperature (Globus, Pascard & Cagan, 1977). However, the most remarkable effect is that substitution of this diamagnetic cation (Zn) results in a significant *increase* in saturation magnetisation in a number of spinel solid solutions, Fig. 4.21. Saturation magnetisation as a function of Zn content shows an increase for small substitutions, goes through a maximum for intermediate values, decreases and finally vanishes for high Zn contents.

A simple analysis shows that this increase can be expected for an antiparallel alignment. As the Zn content increases, magnetic moments decrease in sublattice A and increase in sublattice B. If the magnetic moments of Fe and Ni are 5 and ~ 2.3 μ_B/ion, respectively, the sublattice magnetic moment can be expressed as $5(1-x)$ and that of sublattice B as $2.3(1-x)+5(1+x)$. The magnetisation saturation is given by the

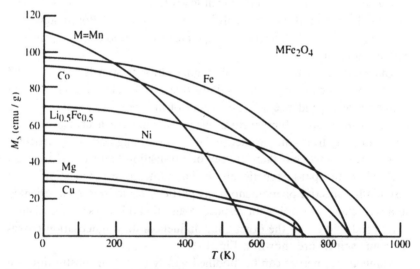

Fig. 4.19. Temperature dependence of the saturation magnetisation for various spinels. Note that M_s is given in emu/g (cgs units). To transform into A/m, magnetisation values have to be multiplied by the density of the corresponding ferrite and then by 10^3. (Adapted from Smit & Wijn, 1961.)

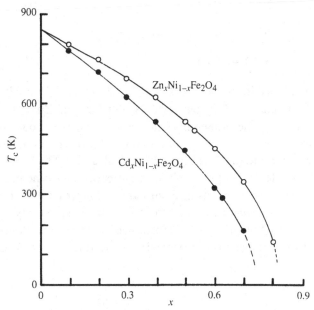

Fig. 4.20. Curie temperature dependence on composition for the
$Zn_xNi_{1-x}Fe_2O_4$ and $Cd_xNi_{1-x}Fe_2O_4$ solid solution series (Globus, Pascard &
Cagan, 1977).

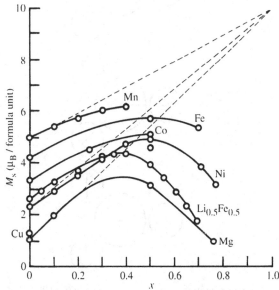

Fig. 4.21. Saturation magnetisation (in Bohr magnetons per formula unit) of various
spinel ferrites at low temperature, as a function of Zn substitution (Gorter, 1954; Guillaud
& Sage, 1951).

difference:

$$M_s(0) = 2.3(1 - x) + 5(1 + x) - 5(1 - x)$$

$$= x(10 - 2.3) + 2.3 \tag{4.41}$$

A linear relationship is obtained with a slope of 7.7, predicting a magnetisation value of 10 μ_B per formula unit for total Zn substitution, $x = 1$, as shown by the broken lines in Fig. 4.21. This relationship is not followed over the entire composition range, however. As Zn content increases, A–O–B interactions become too weak and B–O–B interactions begin to dominate. Instead of a collinear, antiparallel alignment, a *canted* structure appears, where spins in B sites are no longer parallel, Fig. 4.22. Evidence of this *triangular* structure has been observed by neutron diffraction (Satya Murthy *et al.*, 1969); a theoretical analysis showed that departure from collinear order depends on the ratio of the A–O–B to

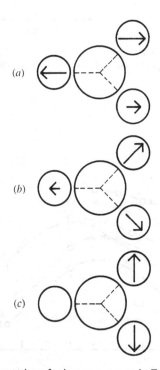

Fig. 4.22. Schematic representation of spin arrangements in $Zn_x Ni_{1-x} Fe_2 O_4$ ferrites: (*a*) ferrimagnetic (for $x \leq 0.4$); (*b*) triangular (or Yafet–Kittel) (for $x \geq 0.5$); and (*c*) antiferromagnetic for $x \approx 1$.

Magnetic order

B–O–B molecular field coefficients, $\lambda_{AB}/\lambda_{BB}$ (Yafet & Kittel, 1952). For high Zn concentration, B–O–B interactions dominate and the ferrites become antiferromagnetic (Boucher, Buhl & Perrin, 1970). A review of configurations of disordered structures has been given by Rao (1989).

Magnetisation in *garnets* presents remarkable features associated with interactions between several magnetic sublattices, such as a compensation temperature and a Curie point virtually independent of the rare-earth cation, Fig. 4.23. Rare-earth garnet sublattices can be represented as $\{\vec{R}_3\}[\vec{Fe}_2](\overleftarrow{Fe}_3)$, where $\{\}$, $[\,]$ and $()$ brackets indicate occupancy of dodecahedral, octahedral and tetrahedral sites, respectively. The arrows show the relative orientation of magnetic moments. The strongest interaction is $[\vec{Fe}_2](\overleftarrow{Fe}_3)$; rare-earth cations are large and coupling with Fe sublattices is quite weak. At low temperatures, the total magnetisation (per formula unit) can be written:

$$M_s(0) = 3\mu_{RE} + 2(5) - 3(5) \tag{4.42}$$

where μ_{RE} is the magnetic moment per rare-earth cation. For $Gd_3Fe_5O_{12}$,

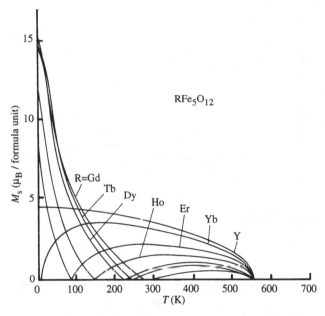

Fig. 4.23. Saturation magnetisation (in Bohr magnetons per formula unit) as a function of temperature, for rare-earth garnets RFe_5O_{12}, where R = Gd, Tb, Dy, Ho, Er, Yb and Y. (Adapted from Moulson & Herbert, 1990.)

133

for example, Gd^{3+} is $4f^7$ and the total moment per ion is taken as $\mu_{RE} \approx 7\mu_B$. Total magnetisation is therefore $21 - 5 = 16\mu_B$. As the temperature increases, the magnetisation in both Fe sublattices decreases slowly; magnetisation in the rare-earth sublattice, in contrast, decreases steeply, Fig. 4.24. For a certain temperature (~ 300 K in GdIG), rare-earth sublattice magnetisation accounts exactly for the difference between tetrahedral and octahedral Fe sublattices, and total magnetisation vanishes. As the temperature increases above the compensation temperature, T_{comp}, the tetrahedral sublattice dominates over the octahedral one, and a net magnetisation reappears, but with an opposite orientation. A similar value for the Curie point for all garnets reflects the fact that the strongest interaction occurs between Fe cations in tetrahedral and octahedral sublattices.

The compensation temperature can be varied by solid solution formation; for example, in $Y_{3-3x}Gd_{3x}Fe_5O_{12}$, virtually any compensation temperature between 0 and 300 K can be obtained simply by adjusting the composition, since $Y_3Fe_5O_{12}$ ('YIG') presents no compensation point (Y is diamagnetic). YIG is an important ferrite; in addition to its use in many

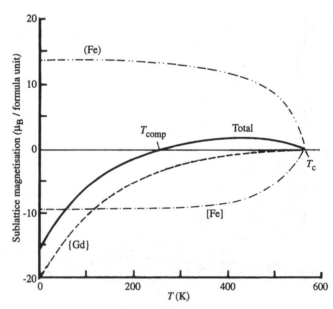

Fig. 4.24. Schematic representation of the sublattice magnetisation dependence with temperature in gadolinium iron garnet (GdIG). Fe in octahedral sites $[Fe^{3+}]$ and Gd in dodecahedral sites $\{Gd^{3+}\}$ are antiparallel to iron in tetrahedral sites (Fe^{3+}).

Table 4.12. *Magnetic moments in hexaferrite building blocks.*

Block	Tetrahedral sites	Octahedral sites	Five-fold sites	Net moment
S	$2_\downarrow$	$3^\uparrow, 1^\uparrow$		$2^\uparrow$
R		$3^\uparrow, 2_\downarrow$	$1^\uparrow$	$2^\uparrow$
T	$2_\downarrow$	$4^\uparrow, 2_\downarrow$		0

Adapted from Smit & Wijn (1961).

applications, it has been used as a 'model' ferrimagnetic material for many fundamental studies.

Magnetisation in *hexagonal* ferrites is complex because their crystal structures are quite complicated. In Ba hexaferrite, $BaFe_{12}O_{19}$, with the simplest structure, Fe cations occupy five different sublattices. As in the description of their crystal structures, it is possible to analyse in detail the superexchange interactions in the building blocks (Smit & Wijn, 1961; Van Loef & Broese van Groenou, 1965; Lilot, Gérard & Grandjean, 1982). In most hexaferrites the strongest interaction is $2f_{vi}$–O–b (or $2f_2$–O–b), between octahedral f_{vi} and five-fold b sublattices (see Section 2.3.1). The expected ordering in the various building blocks is shown in Table 4.12. $BaFe_{12}O_{19}$ is built with RSR*S* blocks; net magnetisation at 0 K is expected to be $M_s = (2^\uparrow + 2^\uparrow + 2^\uparrow + 2^\uparrow)5 = 40\mu_B$ per unit cell, or $20\mu_B$ per formula unit $BaFe_{12}O_{19}$. To transform magnetisation in Bohr magnetons/formula unit to A/m, the following relation is used:

$$M_s = n\mu_B dN_0/W \qquad (4.43)$$

with n = Bohr magnetons per formula, $\mu_B = 9.274 \times 10^{-24}$ A m^2, d = density, 5.28×10^3 kg/m^3, N_0 = Avogadro's number, 6.023×10^{23} g/mol and W = formula weight, 1112 g/mol. Calculation gives $M_s = 5.30 \times 10^5$ A/m, which agrees very well with experimental results, Fig. 4.25. The thermal behaviour of magnetisation is almost linear over a wide temperature range; this has been attributed (Pankhurst *et al.*, 1989) to a rapid decrease in the k sublattice contribution (octahedral Fe^{3+} in the R and S blocks; see Section 2.3.1); deviations from the collinear arrangement appear first in this sublattice, when cation substitution weakens the interactions (Turilli *et al.*, 1986).

Hexagonal ferrites possess a wide range of magnetic properties; Curie

135

Table 4.13. *Curie temperature and saturation magnetisation of some hexagonal ferrites.*

Ferrite	T_C (K)	M_s at 300 K (kA/m)
Ba–M	723	380
Mg_2–Y	553	119
Zn_2–Y	403	227
Fe_2^{2+}–W	728	415
$NiFe^{2+}$–W	793	275
Zn_2–Z	633	310
Cu_2–Z	713	247

Adapted from Smit & Wijn (1961).

Table 4.14. *Saturation magnetisation of various Ba and Sr hexaferrites.*

Ferrite	Magnetisation at 300 K (kA/m)	
	Sr ferrite	Ba ferrite
Fe_2–W	405	415
Ni_2–W	321	335
Co_2–W	399	342
Zn_2–W	429	382
Zn_2–X	393	380
Fe_2–X	377	—

Adapted from Smit & Wijn (1961), Tauber, Megill & Shappiro (1970) and Dey & Valenzuela (1984).

temperature and room-temperature magnetisation data for some compositions are given in Table 4.13. Strontium hexaferrite, $SrFe_{12}O_{19}$ (Sr–M), has a larger magnetisation and coercive field than Ba–M; in Table 4.14, room-temperature magnetisation for various Ba hexaferrites is compared with their Sr counterparts. Zn substitutions have shown a tendency to increase the total magnetisation, presumably by a mechanism similar to that in spinels (Dey & Valenzuela, 1984).

Spontaneous magnetisation is usually *measured* in a strong magnetic field, so that the sample becomes *saturated*, i.e., has only one magnetisation direction. Saturation magnetisation can be determined from the force experienced by the sample in a field gradient. A widely used technique is the 'vibrating sample magnetometer' or VSM; the sample is placed in a long bar, magnetised in a strong dc field, and set to oscillate at a low frequency. The ac field produced by the magnetic flux from the sample is detected by means of a pick-up coil. The amplitude of the coil signal is proportional to the sample magnetisation. This system provides measurements at temperatures from 1 K up to 1000 K, on magnetic moments as small as 5×10^{-8} A m^2 (5×10^{-5} erg/Oe), with an accuracy better than 2%; it was first described by Foner (1959).

4.2.5   Anisotropy and magnetostriction

The magnetisation vector in a crystal is not isotropic, i.e., the total energy of the crystal depends on the orientation of the magnetisation. The directions in which the energy is a minimum are known as *easy directions* and are usually crystallographic axes: $\langle 100 \rangle$ or $\langle 111 \rangle$ in cubic crystals; $\langle 0001 \rangle$ in hexagonal compounds. This phenomenon is known as *magneto-crystalline anisotropy* and is extremely important; it can be said that anisotropy rules all magnetisation processes. The energy needed to deviate the magnetisation vector from an easy direction (usually provided by an applied field) can be represented by a series expansion of the cosines of

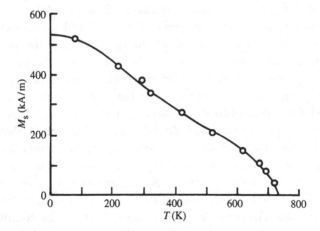

Fig. 4.25. Saturation magnetisation of $BaFe_{12}O_{19}$ as a function of temperature. (Adapted from Smit & Wijn, 1961.)

angles between magnetisation and crystal axes:

$$E_K = K_1(\alpha_1^2\alpha_2^2 + \alpha_2^2\alpha_3^2 + \alpha_3^2\alpha_1^2) + K_2(\alpha_1^2\alpha_2^2\alpha_3^2) + \cdots \qquad (4.44)$$

where K_1 and K_2 are the *magnetocrystalline anisotropy constants*, and the αs are the angle cosines. In some cases, K_2 is negligible, and K_1 is enough to represent anisotropy phenomena. The easy direction determines the sign of K_1. For $\langle 100 \rangle$, K_1 is positive; $\langle 111 \rangle$ leads to negative K_1 values.

In most hexagonal ferrites, the easy direction is the unique crystallographic axis, or *c*-axis. A crystal with only one easy direction is known as a *uniaxial* material, and its anisotropy energy expression is then:

$$E_K = K_1 \sin^2\theta + K_2 \sin^4\theta \qquad (4.45)$$

where θ is the angle of the magnetisation with the *c*-axis. K_1 is positive and K_2 is negative. In some Y-type hexaferrites the easy direction makes an angle of 90° with the *c*-axis; since within the basal plane there are no significant differences in energy as a function of magnetisation direction, this is an *easy basal plane*, resulting in K_1 negative and $K_2 > |K_1|/2$.

The physical origin of magnetocrystalline anisotropy is complex and has not been completely ellucidated. In fact, Eqs. (4.44) and (4.45) are general expressions based on symmetry considerations; all of the phenomenon is included in the anisotropy constants. Magnetocrystalline anisotropy results from spin–orbit interactions. When a spin is deviated from an easy direction by an applied field, the spin–orbit coupling tends to rotate the orbit also, Fig. 4.26. However, orbits are more strongly associated with the lattice, and their rotation needs more energy than the rotation of spins. The 'quenching' of orbital momentum by the crystal field is a manifestation of the orbit–lattice coupling, for example. In many materials, the spin–orbit coupling seems weak enough to be overcome even by small applied fields.

Anisotropy constants vary considerably with temperature. In most cases, anisotropy decreases steeply from a high value at low temperature and then slowly decreases down to zero at T_C. The temperature dependence of the anisotropy constant for $Zn_xNi_{1-x}Fe_2O_4$ ferrites is shown in Fig. 4.27, for various compositions. At low temperatures, total anisotropy can be described in terms of anisotropy contributions from each cation (Kanamori, 1963). The anisotropy of Ni–Zn ferrites at 4 K, for example, can be obtained from contributions of Fe^{3+} in octahedral and tetrahedral sites (Broese van Groenou, Schulkes & Annis, 1967). The contribution from some cations, in particular Co^{2+} and Fe^{2+}, is positive; above a given concentration of any of these cations, the anisotropy constant at low temperatures is positive; however, as the temperature increases, K_1 decreases and eventually becomes negative. This sign change has important implications for magnetisation mechanisms (Section 4.3).

A useful representation of anisotropy is the *anisotropy field*, H_K. Anisotropy effects are described by means of a fictive field, exerting a

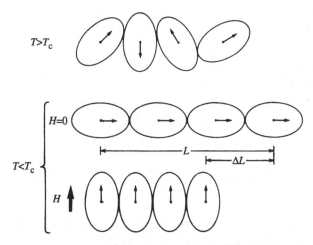

Fig. 4.26. Schematic representation of the origin of magnetocrystalline anisotropy and magnetostriction, in terms of the spin–orbit coupling. (Adapted from Cullity, 1972.)

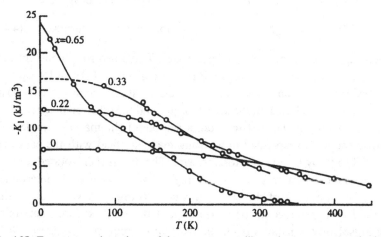

Fig. 4.27. Temperature dependence of the magnetocrystalline anisotropy constant in the Ni–Zn ferrite system for various compositions. (Adapted from Broese van Groenou, Schulkes & Annis, 1967.)

Magnetic properties of ferrites

Table 4.15. *Anisotropy constants and anisotropy fields for various ferrites.*

Ferrite	K_1 (kJ/m^3)	H_K (kA/m)
	(At 300 K)	
Fe_3O_4	-11	24.3
$NiFe_2O_4$	-6.2	24.4
$CoFe_2O_4$	200	500
$MnFe_2O_4$	-3	8
Ba–M	330	1350
Fe_2–W	300	1510
Co_2–Y	-260^a	2230
Zn_2–Y	-100^a	720

^{a} $K_1 + K_2$.
Adapted from Smit & Wijn (1961).

torque equivalent to anisotropy on the magnetic moment. Anisotropy fields, for the various easy directions are as follows:

$$H_K = 2K_1/\mu_0 M_s \qquad \text{for the } \langle 100 \rangle \text{ easy axis;} \qquad (4.46)$$

$$H_K = -4(3K_1 + K_2)/9\mu_0 M_s \qquad \text{for the } \langle 111 \rangle \text{ easy axis;} \qquad (4.47)$$

$$H_K = 2K_1/\mu_0 M_s \qquad \text{for uniaxial materials} \qquad (4.48)$$

where μ_0 is the permeability of free space. The anisotropy constant and anisotropy field for a number of ferrites are given in Table 4.15.

Anisotropy *measurements* require single crystals. There are two basic techniques; one is based on the actual torque, and the other on magnetisation measurements. In the first method, a sample of the material, preferably of disc shape, is suspended in a strong magnetic field with the easy axis parallel to it, by means of a torsion wire. The disc is then rotated and the torque is measured as a function of angle. The value of K_1 is determined from the amplitude of the curves torque $= f$(angle). A variation in this method is to produce an oscillation about the easy axis, and determine the anisotropy constants from the oscillation frequency (Zijlstra, 1967). In the second method, magnetisation curves (magnetisation as a function of applied field) are obtained for the main crystallographic directions. The

Table 4.16. *Magnetostriction*
constant for some ferrites
(*polycrystalline samples*).

Ferrite	λ_s (10^{-6})
Fe_3O_4	$+40$
$MnFe_2O_4$	-5
$CoFe_2O_4$	-110
$NiFe_2O_4$	-26
$Ni_{0.56}Fe^{2+}_{0.44}Fe_2O_4$	0
$Ni_{0.5}Fe^{2+}_{0.6}Fe_2O_4$	-11

Adapted from Moulson & Herbert
(1990).

area under magnetisation curves is related to magnetic energy, and
anisotropy constants can be calculated from them (Cullity, 1972).

Magnetostriction is an effect also related to spin–orbit coupling.
Changes in spin direction result in changes in orbital orientation, slightly
modifying the length of the sample. This is also illustrated in Fig. 4.26. A
common effect of magnetostriction is the humming produced by 50 Hz
transformers. In these devices, an ac current produces strong magnetisation
changes on the core twice a period. Due to magnetostriction, the core
vibrates with a frequency of 100 Hz.

The magnetostriction constant, λ_s, is defined as the strain produced by
a saturating magnetic field, and is dimensionless, $\lambda_s = \Delta L/L_0$, where L_0
is the initial length. A positive magnetostriction means an increase of
length in the field direction. It is a small effect, since λ_s values are usually
$\sim 10^{-5}$. Magnetostriction is anisotropic; length changes in a single crystal
are different for different field orientations. In polycrystals, the observed
λ_s is therefore an average of the single crystal values. Typical values of
the magnetostriction constant for various ferrites are given in Table 4.16.
Usually magnetostriction decreases with temperature before annulating at
T_c.

Magnetisation processes are affected by *stress*. These phenomena, which
can be considered as the opposite of magnetostriction, can also be
visualised on the basis of spin–orbit coupling. The application of a stress
affects the ability of orbitals to rotate under the influence of a magnetic
field. Stress effects can be expressed in terms of a *stress anisotropy*,

141

or magnetoelastic anisotropy, discussed in some more detail in Section 4.3.2.

The interaction of a magnetised solid with its own demagnetisation fields can also be discussed in terms of an additional anisotropy term, *shape anisotropy*. These phenomena, particularly important in small particles, are related to magnetostatic energy and not to spin–orbit coupling. However, demagnetising fields lead to effects similar to the creation of an easy direction for magnetisation. Shape anisotropy is discussed in Section 4.5.1. Since anisotropy and magnetostriction have a common origin, materials with high anisotropy usually also possess a high magnetostriction, but there are exceptions. The *measurement* of magnetostriction is typically performed by associating the change in length with a change in a tuned electric circuit (Zijlstra, 1967).

4.3 Domains and domain walls

4.3.1 Domain structure

As stated earlier, a single crystal of a ferro- or ferrimagnetic material should exhibit a strong magnetic flux on surfaces corresponding to an easy direction, Fig. 4.28(*a*). A parallel order of spins of the whole sample is promoted by exchange, and anisotropy directs them along an easy direction. The crystal is expected to be in a minimum energy state as far as exchange and anisotropy contributions are concerned. Everyday experience, however, shows that a chunk of Fe, for example, is not attracted to another one; no force is experienced between strong ferromagnetic materials in the absence of an external field.

An explanation of this contradiction was first given by P. Weiss in 1906. He assumed that a magnetic material is divided into small volume fractions, called *domains*, which are effectively magnetised to the value of spontaneous magnetisation, but since each domain is oriented along a different easy direction, they compensate each other and the net sample magnetisation is zero. An applied field makes some of the domains grow at the expense of others, thereby resulting in a net magnetisation. *Domain structures* are the most important concept in magnetism; they are the basis of all magnetisation processes.

Domain structure can be understood in terms of the various energies involved. The saturated configuration, i.e., with all spins parallel and oriented along an easy direction, effectively leads to a minimum in exchange and anisotropy energies, E_{ex} and E_K, respectively. However, a magnetic flux outside the sample represents an additional magnetic energy, the *magnetostatic* energy. The discussion is exactly the same for ferro- and for ferrimagnetic materials; the only difference is that instead of magnetic moment per atom, the net resultant per unit cell has to be considered for the latter. The increase in total energy due to magnetic flux generated by sample magnetisation is:

$$E_m = \tfrac{1}{2}N_d M^2 \tag{4.49}$$

where M is the magnetisation and N_d the demagnetisation factor, which depends on the sample's shape. If the sample divides into two domains, Fig. 4.28(*b*), magnetostatic energy decreases significantly, while E_{ex} and E_K contributions still lead to a minimum energy state. Further subdivision results in reductions of E_m. Magnetostatic energy can be virtually eliminated by forming *closure* domains, which retain all magnetic flux within the sample (closure domains are more likely to form in cubic anisotropy). Only a small contribution to the total energy appears in the boundaries between domains.

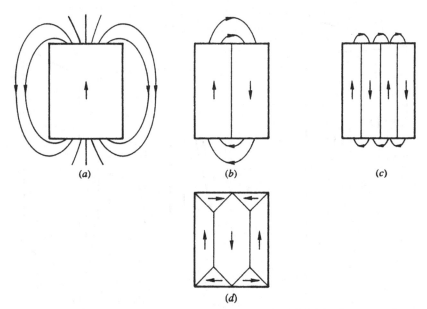

Fig. 4.28. (*a*)–(*c*) Reduction in magnetostatic energy, E_m, by subdivision into magnetic domains. (*d*) Virtual elimination of E_m by creation of closure domains.

Magnetic properties of ferrites

In the boundaries separating one domain from another, known as *domain walls*, magnetisation has to rotate from the direction of one domain (an easy direction) to that of the neighbouring domain (another easy direction); in Fig. 4.29, rotation is by 180°, and these domain walls are known as Bloch walls. To understand this domain wall structure, two extreme wall configurations can be analysed. In the first case, an abrupt rotation is considered; magnetisation experiences an inversion within one interatomic distance, from one domain to the next, Fig. 4.30(*a*). This configuration leads to zero energy contribution from anisotropy, since all spins are oriented along easy directions; however, exchange energy will have a strong contribution from all spin pairs belonging to the domain wall, because each pair contribute, according to Eq. (4.22):

$$E_{ex} = -2J_{ex}s^2 \cos(180°) = +2J_{ex}s^2 \qquad (4.50)$$

Another extreme configuration, Fig. 4.30(*b*), consists of sharing the total rotation angle (180°) by as many spins as possible. In this case, the exchange energy decreases dramatically, since for small angle differences

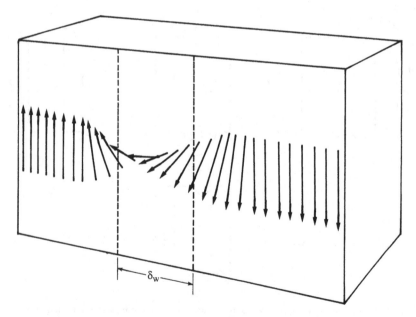

Fig. 4.29. Model of a Bloch domain wall with thickness δ_w. (Adapted from Kittel, 1986.)

between neighbouring pairs, the term $\cos \theta$ can be developed into expansion series ($\cos \theta = 1 - \theta^2/2 + \theta^4/24 + \cdots$) and Eq. (4.22) can be written:

$$E_{ex} \approx -2Js^2 + Js^2\theta^2 \tag{4.51}$$

The first term on the right-hand side is constant and independent of angle and can be eliminated; the second will be small for small angles, $180/N$, where N is the number of spins within the wall. In this hypothetical example, the domain wall includes all the spins from one side to the other of the sample, leading to an exchange energy contribution as small as that in a domain. The only problem with this configuration is that the anisotropy energy is very high; most of the spins are not pointing to easy directions, Fig. 4.30(*b*). Real domain walls have thicknesses which depend on the relative values of exchange (promoting thick walls) and anisotropy (favouring thin walls) energies.

Domain wall energy and thickness for a 180° wall, calculated from the minimum in exchange and anisotropy contributions (Landau & Lifshitz, 1935) can be expressed respectively as:

$$\gamma_w = 2(AK_1)^{1/2} \tag{4.52}$$

$$\delta_w = (A/K_1)^{1/2} \tag{4.53}$$

where γ_w is the domain wall energy per area unit, δ_w the wall thickness, K_1 the anisotropy constant and A the exchange constant, $A = \pi Js^2/a$, and a is the unit cell parameter. To have an idea of the order of magnitude of these parameters, it is interesting to estimate the wall energy and thickness in Fe, at room temperature. Exchange constant A can be roughly taken as $\approx kT_C/a$; $k = 1.38 \times 10^{-23}$ J/K, $T_C = 1043$ K, $a = 2.48 \times 10^{-10}$ m

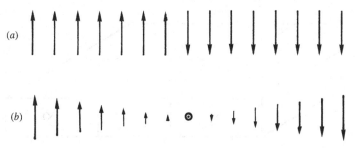

Fig. 4.30. Two extreme configurations of domain wall: (*a*) magnetisation reversal in one interatomic distance; (*b*) magnetisation rotation shared by as many spins as possible.

and $K_1 = 4.8 \times 10^4$ J/m^3 giving:

$$\gamma_w = 2[(1.38 \times 10^{-23})(1043)(4.8 \times 10^4)/(2.48 \times 10^{-10})]^{1/2}$$

$$\gamma_w = 3.34 \times 10^{-3} \text{ J/m}^2$$

$$\delta_w = [(1.38 \times 10^{-23})(1043)/(4.8 \times 10^4)(2.48 \times 10^{-10})]^{1/2}$$

$$\delta_w = 3.48 \times 10^{-8} \text{ m, which is } \sim 140 \text{ unit cells}$$

Domain wall energy in ferrites is expected to be smaller than in metals because T_C and K_1 are smaller and a is larger; wall thickness is smaller as well. Estimates in Ni ferrite lead to $\gamma_w = 6 \times 10^{-4}$ J/m^2 and $\delta_w \sim 60$ unit cells.

A calculation (Cullity, 1972) shows that the total magnetic energy of a crystal (volume ~ 1 cm^3) decreases ~ 1200 times by dividing into ~ 700 domains; further reduction can be expected if closure domains are formed, but an accurate calculation is complex.

Domain structures can be complex in non-uniaxial materials. In addition to 180° walls, closure domain walls in materials with $\langle 111 \rangle$ anisotropy are expected to make angles of 71° and/or 109°, which are the angles between adjacent $\langle 111 \rangle$ directions.

Domain wall structures in thin films and small particles can be different from those in massive samples, because some energy contributions may become significant when sample dimensions are decreased. In thin films, the magnetisation vector tends to remain parallel to the film plane to avoid any contribution to the magnetostatic energy. The spins within a domain wall also rotate within the film plane, which leads to *Néel* walls, Fig. 4.31. Néel walls appear in thin films below a critical thickness limit.

For small particles, there is a *critical* diameter below which the creation of a domain wall results in a larger energy contribution (per volume unit) than the magnetostatic energy of a single domain configuration;

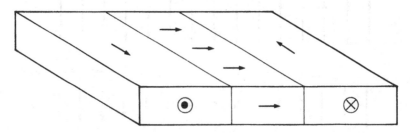

Fig. 4.31. A Néel wall. Spin rotation within the wall occurs without external magnetic flux (schematic).

magnetostatic energy varies as the cubic power of particle diameter, $\sim D^3$, and wall energy depends on $\sim D^2$. Most calculations result in expressions of the form $D = \eta \gamma_w / M_s$, where η is a numerical factor dependent on the assumed geometry. For materials with high anisotropy, these simple calculations result in reasonable agreement with experiment; for low-anisotropy materials, calculated values are usually too small when compared to experiment. Experimental critical diameters are in the 0.1 μm and 1 μm ranges for spinel and hexagonal ferrites, respectively. *Single-domain* particles are the basis of ferrite permanent magnets and other applications such as materials for magnetic recording media.

Domain structures were first observed by the 'Bitter' technique (Bitter, 1931; Williams, Bozorth & Shockley, 1949) in which a colloidal suspension of magnetic particles (usually magnetite) is deposited on the sample's surface; particles are attracted to wall sites because of the flux emerging from wall spins and provide an image of domain structure of the surface. In the Kerr effect, the sample surface is observed with polarised light; domain structure is visualised because magnetisation produces a rotation in the polarisation plane; a sophisticated determination based on this effect allows a quantitative determination of magnetisation (Rave, Schafer & Hubert, 1987). For thin films and slabs (thickness < 5 μm), observation by transmission with polarised light, known as the Faraday effect, provides an image of domains. An example is Fig. 4.32, obtained on YIG.

Transmission electron microscopy can also be used to observe domain walls directly (Hwang *et al.*, 1986). Contrast is obtained by defocusing; electrons passing through one domain are deviated in one direction, and electrons crossing the adjacent domain are deviated in the opposite direction, assuming a 180° wall. A charged particle experiences a force when moving through a magnetic field; this is the Lorentz force, and this technique is therefore known as Lorentz microscopy. A precaution in this method is to reduce or switch off objective lenses; magnetic fields from electromagnetic lenses can affect the original domain structure and even saturate the sample. Domains on sample surfaces can be observed by scanning electron microscopy (Koike & Hayakawa, 1985). Based on magnetostriction, x-ray topography can be used to image magnetic domains in good quality single crystals (Polcarova, 1969). Finally, two new methods promise to give further information on domain structure: magnetic force microscopy (Sáenz *et al.*, 1987) and scanning tunnelling microscopy (Feenstra, 1990). Magnetite has already been studied by the latter (Wiesendanger *et al.*, 1992).

4.3.2   Magnetisation processes and hysteresis

The response of any magnetic material to an applied magnetic field can be understood on the basis of magnetic domains and domain walls. Magnetisation processes are essentially similar in ferromagnetic metals and ferrimagnetic ceramics; the only difference is that in ferrimagnets, instead of a magnetic moment per atom, the resultant of the antiparallel arrangement is considered. The division of a magnetic material into domains explains why, when no field is applied, the magnetic flux is entirely contained within the sample and there is no external manifestation

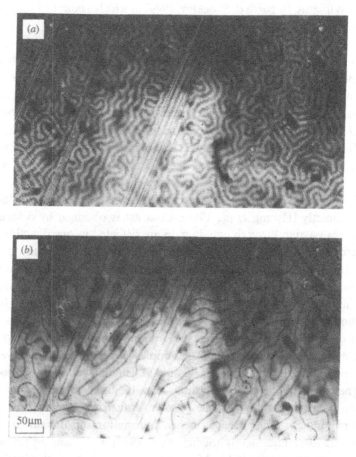

Fig. 4.32. Magnetic domains in a YIG single-crystal platelet, observed by the Faraday effect: (*a*) zero field; (*b*) field normal to the platelet plane ($H \sim 40$ A/m).

of it. The application of an external field, however, can result in a dramatic increase of magnetisation in the sample; an extreme case appears in Fig. 4.33, for a Ni–Fe alloy (Vitrovac®). This *magnetisation curve* shows that a small field of ~ 2 A/m can lead to a magnetisation of ~ 400 kA/m, quite close to its saturation value, $M_s = 477$ kA/m, which represents a susceptibility, M/H, of $\sim 200\,000$. In ferro- and ferrimagnetic materials, it is more common to use the permeability, μ, instead of the susceptibility, because the induction, B, is measured directly. Induction, magnetisation and field are related by:

$$B = \mu_0(H + M) \tag{4.54}$$

Eq. (4.54) is referred to as the *basic equation*. It has a slightly different form in the SI-Kennelly and cgs unit systems (see Tables 4.6 and 4.7). Dividing Eq. (4.54) by H gives:

$$B/H = \mu_0(1 + M/H) \tag{4.55}$$

Since $B/H = \mu$ and $M/H = \chi$, the permeability and susceptibility are related by:

$$\mu = \mu_0(1 + \chi) \tag{4.56}$$

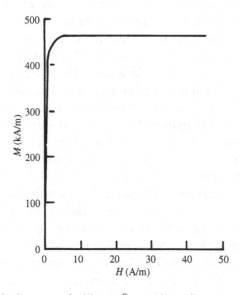

Fig. 4.33. Magnetisation curve of a Vitrovac® amorphous alloy at room temperature.

Magnetic properties of ferrites

The relative permeability, μ_r, is μ/μ_0, which leads to:

$$\mu_r = 1 + \chi \qquad (4.57)$$

Absolute susceptibility and relative permeability are both dimensionless in the SI (Sommerfeld) unit system. The value of the relative permeability gives a good indication of how easily a given material can be magnetised, since $\mu_r = 1$ represents the free space. An additional advantage of using μ_r is that it has exactly the same value in both cgs and SI unit systems. But the important fact in the magnetisation curve of Fig. 4.33 is that a *small* field can lead to an *enormous* magnetisation change! In ferrites, typical permeabilities are smaller; however, values up to $\sim 40\,000$ have been obtained (Beer & Schwarz, 1966; Roess, 1970).

Large permeabilities, i.e., substantially large magnetisation produced by small fields, can only occur because of domain walls. Suppose that a small field, H, with $+z$ orientation is applied to the magnetic material of Fig. 4.29, which has two domains separated by a Bloch wall (one domain with spins oriented along $+z$, parallel to the field, and the other domain along $-z$, opposite to H). Since H is small, it has a negligible effect on the $-z$ spins; these spins are coupled to the $-z$ easy direction by the magnetocrystalline anisotropy. However, the effect of the field on wall spins can be important; these spins are in a sensitive equilibrium state *between* two easy directions (*both* $+z$ and $-z$ are easy directions) and can easily by reoriented by the field. The result is that the wall 'moves' from left to right, increasing the volume of the $+z$ domain at the expense of the $-z$ domain. There is no actual 'displacement' of the wall, but a progressive reversal of spins, as in a wave. However, as in many other collective phenomena, it is easy to visualise it in terms of wall movement. Magnetisation variations are large because even a small wall displacement involves the reversal of spins within a substantial domain volume. Also, the volume change results in a two-fold magnetisation variation since the increase in volume with the field orientation occurs at the expense of domains with the opposite orientation.

The increase in magnetic field leads to domain wall displacements until all the domains with orientations opposite to the field have been substituted by domains with directions parallel to H. However, there might be domains with orientation neither opposite nor parallel; also, the applied field can have an orientation which does not coincide with an easy direction. In the general case, to attain the *saturation* state (in which the sample becomes a single domain oriented along the field direction), a rotation mechanism is required, as indicated in Fig. 4.34. This magnetisation mechanism takes place at high fields; it involves higher energies because the field has to overcome the anisotropy field to produce spin reversal. In contrast, domain wall motion occurs by the progressive reversal of a small fraction of spins, from an easy direction to another one.

An example which illustrates the difference between wall motion and spin rotation mechanisms is the magnetisation of a uniaxial single crystal; for example, Co, Fig. 4.35. Metallic Co has a hexagonal unit cell and its easy directions are parallel to the c-axis. The domain structure is formed by domains with magnetisation oriented along directions $\langle 0001 \rangle$ and $\langle 000\bar{1} \rangle$, separated by Bloch walls. When the field is applied *parallel* to the c-axis, the domain walls are easily displaced until the sample is saturated. A field of $\sim 100 \, \text{kA/m}$ is enough to obtain 98% of the saturation value. The magnetisation curve is quite different when the field is applied *perpendicular* to the c-axis of the crystal, Fig. 4.35. Since domain wall displacements have no effect on the magnetisation in the field direction, the magnetisation process occurs only by spin rotation within each

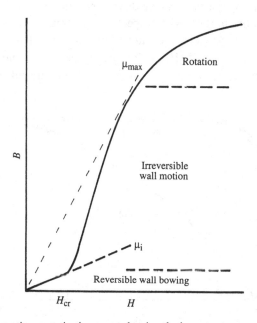

Fig. 4.34. Schematic magnetisation curve showing the important parameters: initial permeability, μ_i (the slope of the curve at low fields), the critical field, H_{cr}, and the main magnetisation mechanism in each magnetisation range.

151

domain. This process requires higher fields than domain motion; saturation of the sample occurs only when the applied field has the value of the anisotropy field, $\sim 800 \text{ kA/m}$.

Magnetisation processes can be affected by stress, Fig. 4.36. These phenomena, known as *stress anisotropy*, are related to magnetostriction and can also be explained on the basis of the spin–orbit coupling, since stress affects the ability of orbitals to be reoriented by the applied field. The sign of the magnetostriction constant determines whether a compressive stress promotes or obstructs the magnetisation process. A negative magnetostriction (magnetisation leading to a reduction in length) results in an increase in magnetisation under compressive stress, as shown in Fig. 4.36 for Ni. A tensile stress leads to a serious decrease in magnetisation. Stress anisotropy affects only the rotational magnetisation mechanism in uniaxial materials; in materials with cubic anisotropy, such as spinel and garnet ferrites, it can also affect the wall motion mechanism. When a Bloch wall is displaced in a uniaxial material, the spins involved in the volume swept are reversed by 180° and the easy axis is the same. In a material with cubic anisotropy, when a 90° wall is displaced, the spins involved in the volume swept by the wall change their orientation by 90° and therefore have a different orientation with respect to the stress axis.

The domain wall displacement is not triggered by any field. There exists a threshold or *critical* field, H_{cr}, below which the wall is not displaced. This critical field depends on sample defects; walls are affected by any

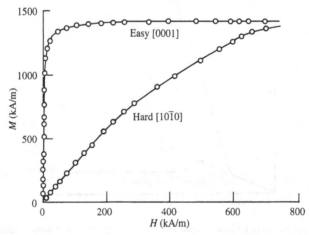

Fig. 4.35. Magnetisation in a Co single crystal with the applied field parallel and perpendicular to the *c*-axis, which is the easy direction. (Adapted from Kaya, 1928.)

deviation from the lattice periodicity: point defects, dislocations, other phases, porosity and even the sample's external surface act as *pinning* sites. The effect of defects can be understood by considering that any discontinuity in magnetic flux represents an additional contribution to the magnetostatic energy. A pore, for example Fig. 4.37(*a*), can give rise to free poles. To prevent the formation of such poles and to reduce magnetostatic energy, closure domains are created, Fig. 4.37(*b*). However, closure domains hinder wall movement. The net result is that defects lead to a higher critical field.

Another important consequence of defects is that even if a wall is displaced, it is pinned in a new position when the field is removed. To obtain a new wall displacement, a larger field than the initial one is usually required. Even during wall displacement, the increase in magnetisation is not completely smooth; if a magnetisation plot is amplified, it can be observed that it is formed by a sequence of steps. These irregularities are due to pinning and unpinning processes and are known as the *Barkhausen* effect.

The critical field appears as a change of slope in the magnetisation

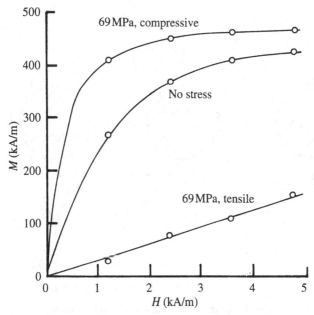

Fig. 4.36. Effects of compressive and tensile stress on the magnetisation curve of Ni. (Adapted from Cullity, 1972.)

curve, Fig. 4.34. For fields lower than the critical value, there is a net magnetisation. This means that even a *pinned* wall can give a response to an applied field. The mechanism responsible for this magnetisation is domain wall *bowing* or *bulging*, Fig. 4.38. A pinned wall can be bent as a flexible membrane under the field 'pressure'; in the energy balance, the increase in total wall energy (due to the increase in wall surface area) is overcome by the decrease in potential energy (as a result of the increase in magnetic moments in the field direction, Eq. (4.12)). An important characteristic of magnetisation produced by wall bowing is that it is *reversible*. Walls recover their planar shape when the field is removed; in the magnetisation plot, Fig. 4.34, field removal leads to zero magnetisation. In addition to reversibility, the permeability for this particular field range tends to be a *linear* function of the field; it is known as the *initial permeability*, μ_i, and is defined as:

$$\mu_i = (\Delta B/\Delta H)_{H\to 0} \tag{4.58}$$

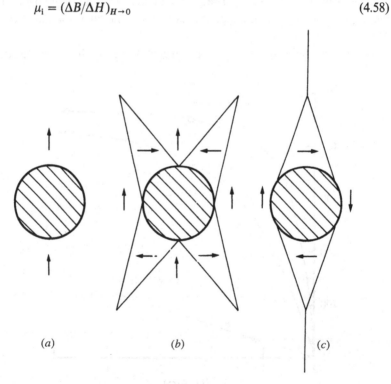

(a) (b) (c)

Fig. 4.37. (a) A pore within a domain leads to free magnetic poles and therefore increases magnetostatic energy. (b) Closure domains are created to reduce free poles. (c) A domain wall on the pore results in a further reduction of poles.

154

The initial permeability corresponds to the slope of the $B = f(H)$ relation at the origin and the critical field to the slope change, respectively, Fig. 4.34 (usually, it is the relative initial permeability, $\mu_i = \Delta B / \mu_0 \Delta H$, that is used). The strong increase in magnetisation at fields larger than the critical value is therefore a consequence of wall unpinning and displacement. Wall displacement 'sweeps' volumes significantly larger than wall bowing and leads therefore to a greater slope in the $M(H)$ plot.

A plot of permeability as a function of field is often useful to determine characteristics such as initial permeability, μ_i, and *maximum* permeability, μ_{max}, Fig. 4.39. This plot represents the slope of the $B(H)$ relationship, as a function of H. The initial part at low fields is the initial permeability, μ_i, and the maximum corresponds therefore to μ_{max}, Fig. 4.34. As H increases

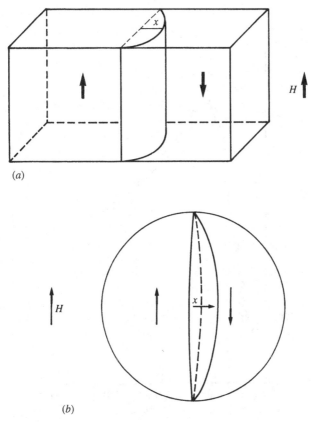

(a)

(b)

Fig. 4.38. Magnetisation by reversible domain wall bowing: (a) a rectangular wall pinned on two edges; (b) a circular wall in a spherical grain, pinned on its perimeter.

above this value, magnetisation approaches saturation and permeability decreases as $\sim 1/H$.

The removal of a strong field (larger than the critical value) does not result in zero magnetisation, but in a *remanent* state, B_r, Fig. 4.40. Domain walls are pinned in a new position, in which some net domain volume remains in the original field direction. A new wall displacement requires a new critical field, now in the opposite direction and usually larger than the field value for the initial displacement from the zero magnetisation state. Instead of critical field, this field is known as *coercive* field, coercive force or coercivity, H_c. When the value of this field is overcome, a reversal of magnetisation is produced and a tendency to saturation in the new orientation appears, Fig. 4.40. Removal of the field leads to the symmetric remanent magnetisation. A cyclic field results therefore in a loop; this is

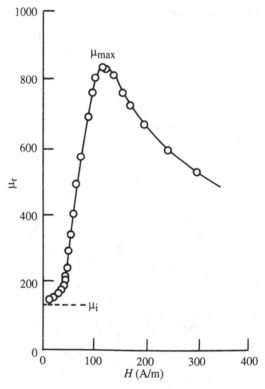

Fig. 4.39. Magnetic permeability of YIG as a function of applied field: the maximum value is μ_{max} (Globus, Duplex & Guyot, 1971).

the *hysteresis* loop, which is in some way a synthesis of almost all magnetic phenomena.

In polycrystalline materials, the hysteresis loop can be described by a combination of bowing, unpinning and displacement processes on grain boundaries (Globus, 1962, 1977). Starting from the demagnetised state, Fig. 4.41, the sample is represented by a single grain divided into two domains, separated by a diametral Bloch wall. Application of a small field leads to reversible wall bowing, corresponding to initial permeability. For $H > H_{cr}$, the wall is unpinned and displaced within the grain, resulting in a strong increase in magnetisation. Removal of the field does not lead to elimination of magnetisation, because the wall is now pinned in a new position within the grain. The small reduction in magnetisation towards the remanent state is due to the fact that the wall recovers its planar shape. A field with the opposite orientation is now applied. Wall bowing occurs for small fields; the critical field value for this new position is generally larger than for the diametral position. When this value is achieved, the wall is unpinned and displaced towards the other end of the grain, where it is pinned again on field removal.

On the basis of this model, a quantitative description of magnetisation

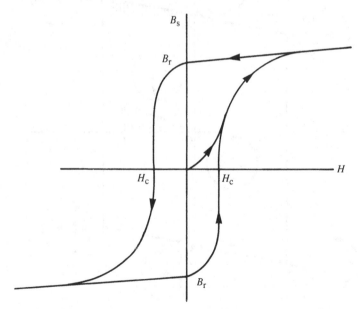

Fig. 4.40. Schematic hysteresis loop. B_s and B_r are the saturation and remanent inductions, respectively; the coercive field is H_c.

processes has been made (Escobar, Magaña & Valenzuela, 1983), including the effects of grain size distribution (Magaña, Escobar & Valenzuela, 1986), in good agreement with experimental results. The coercive field varies *inversely* with grain size, Fig. 4.42; the width of a hysteresis loop can therefore be controlled within certain limits by controlling the grain size. By taking into account all the parameters affecting hysteresis loops, Globus (1977) has proposed a 'universal hysteresis loop', which includes hysteresis loops of a number of ferrite compositions, measured over a wide temperature range, Fig. 4.43. The hysteresis loops were normalised by plotting M/M_s as a function of HD_m/H_{eff}, where D_m is the average grain size and H_{eff} is the total anisotropy field, including magnetocrystalline and stress anisotropies (Globus, Duplex & Guyot, 1971).

During the generation of a hysteresis loop, the walls are displaced back and forth; in an idealised spherical grain such as the one in Fig. 4.41, these displacements involve the destruction and creation of a fraction of the

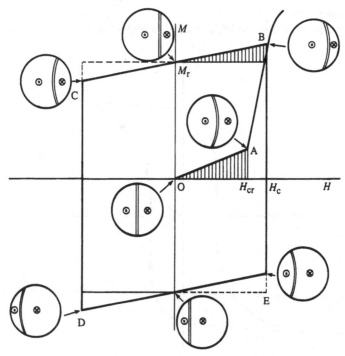

Fig. 4.41. Magnetisation curve OAB, and hysteresis loop BCDEB as combinations of domain wall bowing, unpinning and displacement (Globus, 1977).

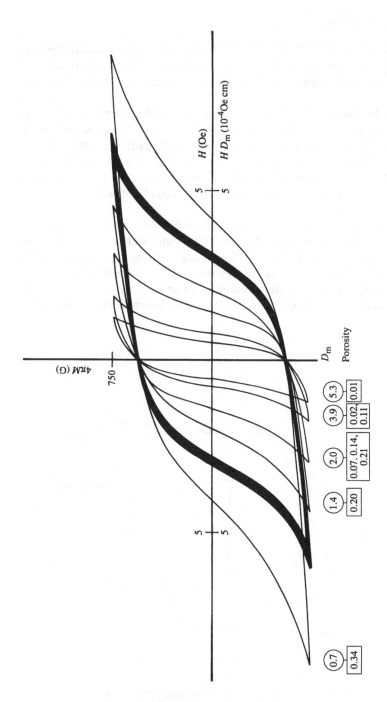

Fig. 4.42. Effect of average grain size, D_m, on the coercive field of YIG. A single loop is obtained by plotting the magnetisation as a function of HD_m (Globus, 1977).

Magnetic properties of ferrites

total wall surface area. Guyot & Globus (1977) showed that this wall energy (the summation of the creation *and* destruction of part of the wall) was accurately represented by the area contained by the hysteresis loop.

In addition to domain wall bowing and displacement, there are two other magnetisation mechanisms: *spin rotation* and *spin waves*. In the former, Fig. 4.44, magnetic moments within domains are simply rotated out of their easy direction by the external field. This mechanism occurs for virtually any magnitude of field; the rotational permeability is a linear function of the field; it is reversible and is typically small. It is discussed in Section 4.5.1.

Spin waves or *magnons* are an additional manifestation of the quantum-mechanical nature of ordered spins. The ground state of a ferromagnetic spin system at 0 K is the parallel arrangement of magnetic moments, Fig. 4.45(*a*). An excited state can be represented by the reversal of one spin, Fig. 4.45(*b*). However, the precession of spins sharing a small angle between them represents an excited state with lower energy (Kittel, 1986).

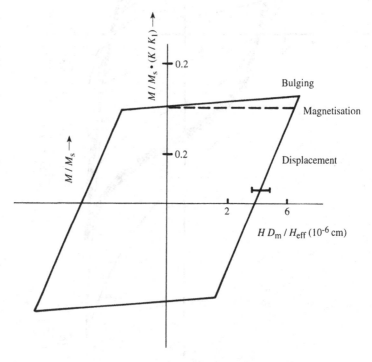

Fig. 4.43. Universal hysteresis loop, obtained by plotting M/M_s as a function of HD_m/H_{eff}, where H_{eff} is the effective anisotropy field (Globus, 1977).

Domains and domain walls

A wave can be propagated through the material by means of this mechanism, Fig. 4.45(c). In a similar way to phonons (quantised mechanical vibrations of the lattice), magnons can be thought of as particles, with all their associated properties and interactions. Spin wave theory (Bloch, 1930) provides an interpretation of the low-temperature behaviour of saturation magnetisation in ferromagnets.

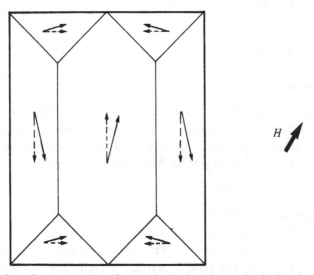

Fig. 4.44. Schematic representation of magnetisation by spin rotation in domains.

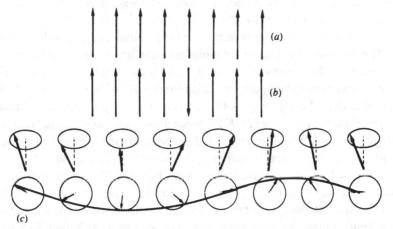

Fig. 4.45. Magnetisation by spin waves: (a) ground state; (b) excited state by inversion of a spin; (c) excited state by spin precession with a small phase lag, creating a spin wave. The spin wave mechanism possesses a lower total energy than the spin inversion in (b).

161

Magnetic properties of ferrites

A magnetic material which is easy to magnetise (and demagnetise) is known as a *soft* material, and permanent magnets are referred to as *hard* magnets. The boundary in coercive field value for these two cases is somewhere between $1\,\text{kA/m}$ (soft if $H_c \leq 1\,\text{kA/m}$) and $10\,\text{kA/m}$ (hard for $H_c \geq 10\,\text{kA/m}$). In ferrites, Mn–Zn spinels have the lowest values ($H_c \sim 16\,\text{A/m}$) and $BaFe_{12}O_{19}$ particles the highest ($H_c \sim 420\,\text{kA/m}$, Mee & Jeschke (1963)).

4.4 Soft ferrites

4.4.1 Initial permeability

An important property of soft magnetic materials is the initial permeability. Its dependence on microstructure can be understood in terms of the Globus model. A calculation of the (relative) initial permeability (Globus, Duplex & Guyot, 1971) leads to the following expression:

$$\mu_i \approx 3\mu_0 M_s^2 D/16\gamma_w \qquad (4.59)$$

where μ_0 is the permeability of free space, M_s is the saturation magnetisation (in A/m), D is the average grain size (in m) and γ_w is the domain wall energy (in J/m^2). The initial permeability is therefore a linear function of grain size, Fig. 4.46. The value at zero grain size represents the *rotational* permeability. Porosity and other defects such as precipitates do not affect the initial permeability if they are confined to grain boundaries (intergranular defects); otherwise, the initial permeability is severely decreased since defects act as pinning sites, reducing the volume swept by wall bowing.

The thermal behaviour of the initial permeability is important for many applications. In a representative spinel solid solution series, Ni–Zn ferrites, the relative permeability can be varied from ~ 20 for $NiFe_2O_4$, to ~ 8000 for $Zn_{0.7}Ni_{0.3}Fe_2O_4$, Fig. 4.47. As stated by Eq. (4.59), the initial permeability varies as $\approx M_s^2/K_1^{1/2}$. Since anisotropy decreases faster than magnetisation on heating, the initial permeability is expected to increase with temperature, tend to infinity just below the Curie point and then drop to ≈ 1 for the paramagnetic phase, Fig. 4.48. The peak near T_C is known as the 'Hopkinson' peak. Co^{2+} and Fe^{2+} cations in ferrites have a positive contribution to anisotropy; this contribution can dominate at low temperatures. As the temperature increases, this contribution decreases and anisotropy eventually becomes negative. Ferrites containing one of these cations exhibit a second permeability maximum, due to the change

162

of sign in K_1, Fig. 4.49. Variation in composition results in different temperatures for the secondary peak, Fig. 4.50.

Measurement of the initial permeability as a function of temperature can therefore be used as a material characterisation method. The value of T_C depends only on the composition; the verticality of the permeability drop at the Curie point indicates the degree of homogeneity in the sample composition (Cedillo *et al.*, 1980; Valenzuela, 1980).

4.4.2   *Disaccommodation and magnetic annealing*

In some ferrites, the initial permeability decreases as a function of time. This phenomenon, known as *disaccommodation*, is more apparent when

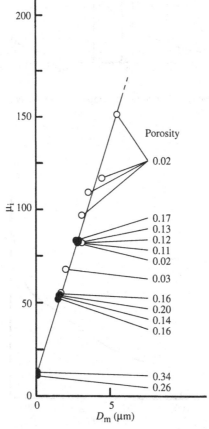

Fig. 4.46. Dependence of the initial permeability on average grain size in YIG: intergranular porosity values are indicated (Globus & Duplex, 1966).

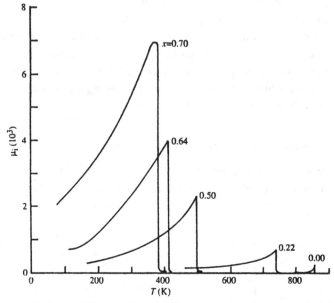

Fig. 4.47. Initial permeability as a function of temperature in Zn_xNi_{1-x} spinels for various compositions.

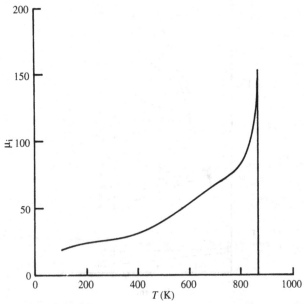

Fig. 4.48. The initial permeability of Ni ferrite as a function of temperature (Cedillo *et al.*, 1980).

a strong, ac field (larger than the critical field) is applied for a few seconds prior to the observation. The decrease in permeability is larger at higher temperatures, Fig. 4.51. Disaccommodation is due to interactions between point defects and domain wall pinning. Consider a magnetic material with point defects such as interstitials or vacancies, at a temperature high enough so that diffusion is not negligible. Due to the spin–lattice coupling

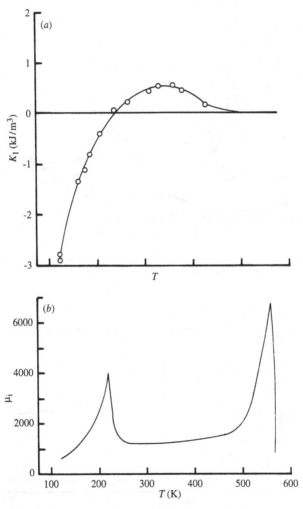

Fig. 4.49. (*a*) Magnetocrystalline anisotropy constant and (*b*) initial permeability of a $Zn_{0.2}Mn_{0.6}Fe_{2.2}O_4$ single crystal. The permeability maxima correspond to zero values of K_1. (Adapted from Ohta, 1963.)

(see Section 4.2.5), defect distribution can lead to a slightly different total energy depending on its arrangement with respect to the particular easy direction in that domain. A *directional* order can be established, in which cation pairs are aligned, for example, parallel to the magnetisation direction. A non-negligible concentration of point defects (essentially cation vacancies) is required to promote diffusion. In turn, directional

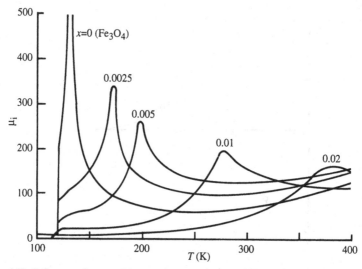

Fig. 4.50. Influence of composition on the second permeability maximum in $Co_xFe_{3-x}O_4$ ferrites (Bickford, Pappis & Stull, 1955).

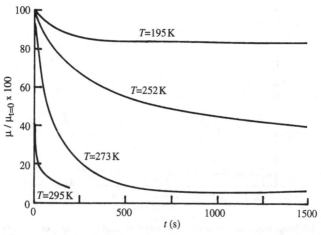

Fig. 4.51. Disaccommodation in a $Mn_{0.85}Fe_{2.15}O_4$ single crystal. The initial permeability was normalised to 100 for $t = 0$ at each temperature. (Adapted from Enz, 1958.)

ordering results in an *induced anisotropy* which adds to the magneto-crystalline anisotropy.

The effects of defects can be more dramatic on walls than in domains. In a domain wall, magnetisation vectors are progressively rotated by a small angle within the wall thickness; as in domains, defects or cation pairs are attracted to wall sites to decrease the total energy by creating directional order arrangements. Since spin orientation varies substantially within the small volume of a wall, defect localisation leads to complex configurations. This process results in an increased wall pinning. Disaccommodation can be explained as follows. The strong field prior to initial permeability measurements unpins walls and destroys the directional order established at wall sites. When this field is removed, walls are pinned again at grain boundaries, since the measuring field is small (it is lower than the critical value). Formation of cation pairs (or localisation of defects) takes place on wall sites, with a rate depending on the defect diffusion kinetics. In ferrites containing Co^{2+}, the effects are attributed to directional ordering of cobalt cations in $\langle 111 \rangle$ directions; in Mn–Zn ferrites, disaccommodation is attributed to directional ordering of Mn^{2+}–Fe^{2+} pairs (Cullity, 1972).

Induced anisotropy phenomena also provides a technique for increasing the initial permeability by means of *magnetic annealing*. This method requires negligible defect diffusion at room temperature. The first step is to eliminate the complex directional ordering established in wall sites, which can be done by a thermal treatment (or annealing), typically 1–4 hours at a temperature high enough to promote diffusion, under a saturating dc field. Since there are no walls, pair ordering occurs homogeneously in the whole sample with the orientation imposed by the external field. Still under field, the sample is then cooled to room temperature. A variation in this method is to anneal the sample at a temperature above T_C with no field; directional order is eliminated in the paramagnetic phase. The cooling rate down to room temperature has to be rather fast, to avoid the formation of directional ordering during cooling.

4.5   Hard ferrites

4.5.1   Magnetisation rotation

Permanent magnets, made of hard ferrites, are used in applications where a constant magnetic field without electric current is needed. A permanent magnet possesses therefore *stored* energy, which was provided during material preparation, in particular, during magnetisation of the sample. Since the aim is to produce the highest possible field, hard ferrites with high saturation magnetisation are used. The difference between remanent and saturation magnetisation depends on the microstructure. The stability of the material when subject to external fields is also very important; a high coercivity is therefore required.

The effect of grain size on coercive field is shown in Fig. 4.52 for several ferrites. The reduction in grain size results in single-domain grains and a maximum in the coercive field; in the absence of domain walls, the main magnetisation mechanism is *rotation* of spins within domains. The magnetisation vector is 'pulled' out of the easy direction by the external field, working directly against the anisotropy field, H_K (see Section 4.2.5).

When a magnetic field is applied to a single-domain sample (with uniaxial anisotropy) at a 90° angle, no hysteresis loop is obtained. The field deviates the magnetisation from its easy direction and leads to saturation for $H \geq H_K$, but the magnetisation recovers (in a reversible way) its easy direction when the field is removed, Fig. 4.53. The other interesting case is when the field is applied along the easy direction. A rectangular hysteresis loop is obtained, Fig. 4.53; magnetisation is abruptly reversed when the field attains the value H_K. The hysteresis loop of a

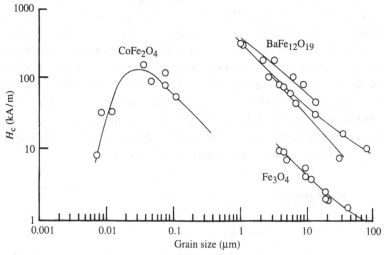

Fig. 4.52. Dependence of the coercive field on grain size for some ferrites. (Adapted from Luborsky, 1961.)

168

polycrystalline sample formed with single-domain, uniaxial particles with random orientation was calculated by Stoner and Wohlfarth (1948), Fig. 4.54. Differences with experimental results arise because reversal of magnetisation can occur by other mechanisms (such as domain nucleation, or wall unpinning in particles large enough to form domains); theoretical values are always greater by a factor of 2–3. Another source of disagreement is that this theoretical model assumes no interaction between particles, which is quite far from the actual conditions in high-density assemblies of particles. An accurate calculation, however, is extremely complex.

In addition to magnetocrystalline anisotropy, single-domain particles exhibit a *shape anisotropy*. The origin of this anisotropy can be visualised by considering the difference in magnetostatic energy in a non-spherical particle, Fig. 4.55. The demagnetising field is stronger for a short axis than for a long one. For a prolate spheroid ($\sim$rod), the shape-anisotropy

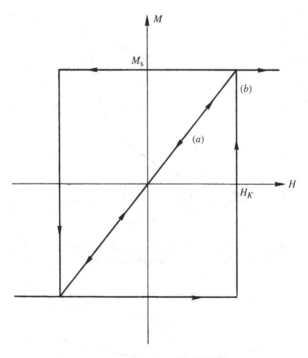

Fig. 4.53. Magnetisation behaviour of a single-domain particle assuming rotation-only mechanism: (*a*) applied field at 90° from the easy direction; (*b*) field collinear to the easy direction (Stoner & Wohlfarth, 1948).

Magnetic properties of ferrites

constant is given by:

$$K_s = (N_a - N_c)M^2/2 \qquad (4.60)$$

where K_s is the shape-anisotropy constant, N_a and N_c are the demagnetising coefficients along axes a and c, respectively (see Cullity (1972)), and M is the particle magnetisation. In a rod, for instance, the magnetisation vector therefore tends to remain within the long axis. If magnetocrystalline anisotropy coincides with the long axis both anisotropies add; otherwise, the easy direction is dominated by the stronger anisotropy. For some single-domain particle applications (see Chapter 5), the availability of needle-shaped particles is therefore more important than their intrinsic anisotropy.

Ba hexaferrite particles usually grow as plates with their c-axis perpendicular to the plate surface. This growth habit leads to a shape anisotropy perpendicular to the magnetocrystalline anisotropy and therefore to a

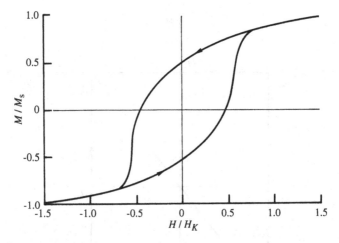

Fig. 4.54. Theoretical hysteresis loop of a polycrystalline sample made of single-domain particles with random orientation of easy axes (Stoner & Wohlfarth, 1948).

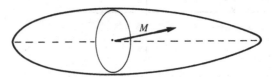

Fig. 4.55. Origin of the shape anisotropy. The magnetisation tends to remain oriented along the longest axis (schematic).

170

reduced total anisotropy field. The total anisotropy field is given by (Mee & Jeschke, 1963):

$$H_K = 2K/\mu_0 M_s - N_d M_s \qquad (4.61)$$

where the reduction in magnetocrystalline anisotropy due to shape anisotropy is $N_d M_s$; N_d is the demagnetisation factor, which for thin plates is 1. The reduction in anisotropy is (Tables 4.13 and 4.15) from 1350 kA/m to 970 kA/m.

4.5.2   The $(BH)_{max}$ product

The magnetisation process for storing magnetic energy in a hard material appears in Fig. 4.56. It is usually represented in the form of B vs H plots, because the induction is measured directly. After application of a strong field to magnetise the sample, the field is removed and induction decreases to the remanent value, B_r. Permanent magnets are always shaped to produce an external magnetic flux; they are said to work in *open circuit*. As a result, a demagnetising field, H_d, appears leading to a further decrease of induction from the remanence to a value B_d, on the *demagnetisation* curve. The shape of the demagnetising curve determines the energy stored, which is directly proportional to the product BH, Fig. 4.57. Permanent

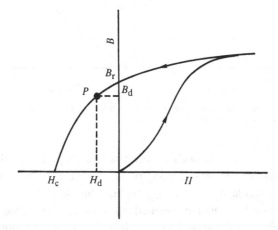

Fig. 4.56. Irreversible magnetisation process. After application and removal of a strong field, the sample is in the remanence, B_r; however, due to the demagnetising field, H_d, induction decreases to the B_d value. Operation point is P.

Table 4.17. *Remanence, coercivity and energy product of some hard ferrites.*

Ferrite	B_r (T)	H_c (kA/m)	$(BH)_{max}$ (kJ/m^3)
Ba–M isotropic	0.225	147	9.55
Ba–M anisotropic	0.360	230	24.7
Sr–M anisotropic	0.343	265	23.1
Fe^{2+}–W	0.470	167	34.2

Adapted from Cullity (1972) and Lotgering, Vromans & Huybert (1980).

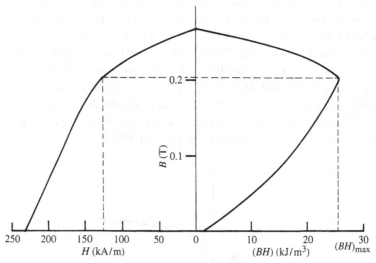

Fig. 4.57. Schematic determination of the maximum energy product from the demagnetisation curve.

magnets are designed in such a way that its shape leads (through the demagnetising field) to an operating point P in the demagnetisation curve, as close as possible to the $(BH)_{max}$ value, Fig. 4.57. Hard magnets are usually evaluated by their demagnetisation curves, presented on special plots which include constant $(BH)_{max}$ lines, Fig. 4.58. The units of energy product are kJ/m^3 in SI and MGOe in cgs; 1 MGOe = 7.96 kJ/m^3. Some representative values for hard ferrites are given in Table 4.17.

4.6   *Magnetisation dynamics*

4.6.1   *Domain wall dynamics*

Many of the specific applications of ferrites depend on their behaviour at high frequencies. When subjected to an ac field, ferrite permeability shows several *dispersions*; as the field frequency increases, the various magnetisation mechanisms become unable to follow the field. The dispersion frequency for each mechanism is different, since they have different *time constants*, Fig. 4.59. The low-frequency dispersions are associated with domain wall dynamics and the high-frequency dispersion, with spin resonance; the latter, usually in the GHz range, is discussed in Section 4.6.2.

The two main magnetisation mechanisms are wall bowing and wall displacement (see Section 4.3.2); in fact, any field results in a bowing of pinned walls, and if this field has a higher value than the corresponding critical field, walls are unpinned and displaced. Otherwise, bowed walls remain pinned to material defects. Measurements at low fields therefore show only one wall dispersion, Fig. 4.60; at high fields, several, complex dispersions are observed, such as those in Fig. 4.59. Wall displacement

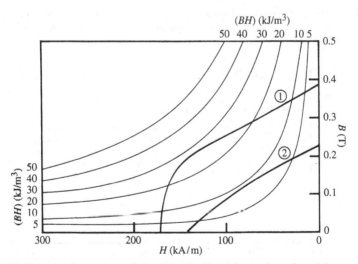

Fig. 4.58. Demagnetisation curve for anisotropic (1) (with a grain preferential orientation) and isotropic (2) samples of $BaFe_2O_4$. Constant (BH) lines are shown (schematic).

173

dispersion occurs at lower frequencies than wall bowing, since hysteresis is a more complex phenomenon involving a combination of wall bowing, unpinning, displacement and pinning steps.

Domain wall dynamics are usually represented by an *equation of*

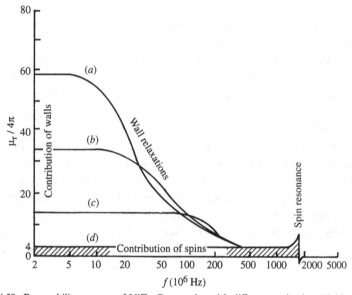

Fig. 4.59. Permeability spectra of $NiFe_2O_4$ samples with different grain size: (*a*) 11 μm; (*b*) 5 μm; (*c*) 2 μm; and (*d*) size < 0.2 μm (single-domain behaviour) (Globus, 1962).

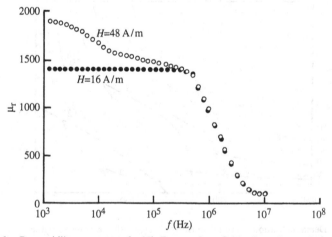

Fig. 4.60. Permeability spectrum of a Ni–Zn sample at fields above (open circles) and below (filled circles) the critical field.

Magnetisation dynamics

motion:

$$m \, \mathrm{d}^2x/\mathrm{d}t^2 + \beta \, \mathrm{d}x/\mathrm{d}t + \alpha x = 2M_\mathrm{s} H(t) \qquad (4.62)$$

where m is the effective wall mass, β the viscous damping factor, α the restoring constant, x the wall displacement and $H(t)$ the driving field; m, β and α are defined on the basis of the wall area unit. The first term on the left-hand side represents the wall inertia as a product of mass times the acceleration; the second is the damping opposing the propagation velocity, and the third is associated with wall pinning to defects, expressed as a restoring force. In experiments using sinusoidal fields, results can be analysed by using the *complex* susceptibility $\chi^* = \chi' - \mathrm{j}\chi''$, where χ' and χ'' are the real and the imaginary components of the complex permeability and $\mathrm{j} = (-1)^{1/2}$. They can be expressed as (Vella-Coleiro, Smith & Van Uitert, 1972):

$$\chi' = \chi_0 \frac{1 - (\omega/\omega_\mathrm{s})^2}{(1 - \omega^2/\omega_\mathrm{s}^2)^2 + (\omega/\omega_\mathrm{x})^2} \qquad (4.63)$$

$$\chi'' = \chi_0 \frac{\omega/\omega_\mathrm{x}}{(1 - \omega^2/\omega_\mathrm{s}^2)^2 + (\omega/\omega_\mathrm{x})^2} \qquad (4.64)$$

where χ_0 is the low-frequency value of susceptibility, ω is the angular frequency of the driving field ($\omega = 2\pi f$) and ω_s and ω_x are the resonance and relaxation frequencies, respectively. For high permeability materials, $\mu_0 \gg 1$, $\mu_0 \approx \chi_0$ and the real and imaginary permeabilities can be expressed by identical equations to (4.63) and (4.64), respectively. Resonance and relaxation frequencies are related to effective mass and damping and restoring force factors by:

$$\omega_\mathrm{s} = (\alpha/m)^{1/2}; \qquad \omega_\mathrm{x} = \alpha/\beta \qquad (4.65)$$

Damping effects usually dominate inertial effects ($\beta > m$) and bowed wall dispersion shows a relaxation behaviour, as in Fig. 4.60. In a pinned wall, the relaxation frequency depends on the grain size, D, as $\sim 1/D^2$ (Gieraltowski & Globus, 1977) and the initial permeability varies as $\sim D$; this means that permeability and relaxation frequency are not independent, but are related by $\mu_0^2 \omega_\mathrm{x} = \text{constant}$. Domain wall resonance has been observed in some garnet single crystals (Vella-Coleiro *et al.*, 1972) as well as in polycrystalline garnets and spinels (Gieraltowski, 1989). The maximum in the resonance peak was found to be related to deviations in the grain-size distribution.

175

The techniques of impedance spectroscopy, widely used in dielectrics (Jonscher, 1983; MacDonald, 1987) have been applied to magnetic materials. In this method, impedance measurements as a function of frequency are modelled by means of an equivalent circuit and its elements are associated with the physical parameters of the material. The complex permeability, μ^*, is determined from the complex impedance, Z^*, by:

$$\mu^* = (jk/\omega)Z^* \tag{4.66}$$

where k is the geometric constant relating inductance, L, to the permeability. The equivalent circuit for domain wall bowing (applied field lower than critical field) is a parallel RL arrangement; for wall displacement, an additional Warburg-type impedance element is required (Irvine *et al.*, 1990a).

A different type of experiment can be performed on single crystals to measure the domain wall propagation at *constant* velocity, by applying a constant field. The acceleration term is eliminated and Eq. (4.62) can be written:

$$v = C(H - H_p) \tag{4.67}$$

where v is the wall velocity, C is the domain wall mobility $(=2M_s/\beta)$ and H_p, known as the propagation field, is the critical field, Fig. 4.61. The damping factor depends on two mechanisms: *intrinsic* damping and *eddy-current* damping. Intrinsic damping, which occurs in all materials,

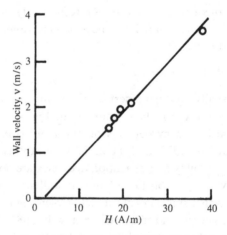

Fig. 4.61. Domain wall velocity as a function of applied field for a Mn–Zn ferrite single crystal at 77 K. (Adapted from Dillon & Earl, 1959.)

is associated with a limiting velocity in spin reversal during wall propagation. Eddy-current damping represents the hindering of wall displacement by the generation of eddy currents and therefore depends on electrical resistivity. Since ferrites have high resistivities, their wall mobility is very high ($\sim$3.8 m/(s A/m) for Ni ferrite (Galt, 1954)) as compared with metals ($\sim$7 $\times$ 10^{-4} m/(s A/m) in Si–Fe (Williams, Shockley and Kittel, 1950)).

4.6.2 Ferromagnetic resonance

At very high frequencies, domain walls are unable to follow the field and the only remaining magnetisation mechanism is spin rotation within domains. This mechanism eventually also shows a dispersion, which always takes the form of a *resonance*. Spins are subjected to the anisotropy field, representing spin–lattice coupling; as an external field is applied (out of the spins' easy direction), spins experience a torque. However, the response of spins is not instantaneous; spins *precess* around the field direction for a certain time (the relaxation time, τ) before adopting the new orientation, Fig. 4.62. The frequency of this precession is given by the Larmor frequency:

$$\omega_L = \gamma \mu_0 H_T \qquad (4.68)$$

where H_T is the total field acting on the spin, $H_T = H_K + H + H_d + \dots$, where H_K, H, and H_d are the anisotropy and the external and demagnetisation

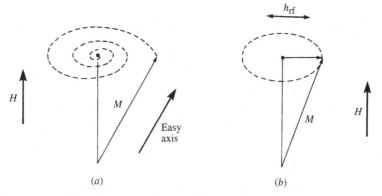

Fig. 4.62. (*a*) Schematic representation of the spin deviation from an easy axis by precessional spiralling into the field direction. (*b*) Precession is maintained by a perpendicular rf field, h_{rf}. (Adapted from Moulson & Herbert, 1990.)

fields, respectively. If an ac field of angular frequency ω_L is applied to the sample, a resonant absorption (ferromagnetic resonance) occurs. The Larmor frequency is independent of the precession amplitude.

Ferromagnetic resonance experiments are usually performed by saturating the sample with a strong dc field and applying a small, perpendicular ac field at a constant frequency. The dc field is then slowly varied to achieve resonance conditions, measuring the power absorption in the sample. This is the best arrangement for resonance experiments since all the spins are oriented in a given direction in the saturated sample, and the ac field exerts a torque on their perpendicular component. Experimentally, it is easier to use an ac field of high, constant frequency (usually in the tens of GHz range) and slowly vary a strong dc field to achieve resonance conditions, than to use other combinations. Resonance is therefore plotted as power absorption as a function of dc field, for a given, constant frequency, Fig. 4.63. Ferromagnetic resonance experiments provide a method of determining anisotropy constants through measurement of anisotropy fields.

The linewidth of resonance peaks (width measured at half the height of resonance peak), ΔH, depends on a variety of factors. Anisotropy field effects can be expressed in terms of a damping constant leading to a contribution to ΔH. If the dc field is not strong enough to saturate the

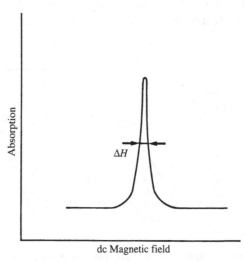

Fig. 4.63. Ferromagnetic resonance. Larmor conditions are achieved with an ac field of constant frequency and a strong, slowly varying, dc field; ΔH is the resonance linewidth.

sample, the total field in each domain is different and ΔH increases. In polycrystalline samples this effect is more serious, because interactions between grains result in larger variations in the total field. Local perturbations in H_T produced by porosity lead to a linear increase in linewidth. Finally, even surface roughness can produce significant contributions to ΔH.

4.7   Electrical properties

Electrically, ferrites can be classified as somewhere between semiconductors and insulators. In many applications, this is their main advantage over ferromagnetic metals, because their high resistivity results in low energy losses. When an ac field is applied to a conductive material, the fraction of the field absorbed to excite the conduction electrons becomes increasingly important as the frequency increases, at the expense of the field fraction used to magnetise the sample effectively. An accurate calculation of eddy-current losses is extremely complex, because it depends on the detailed domain structure. However, an approximate comparison can be illustrative. Eddy-current losses can be expressed as:

$$P_e = KB^2f^2d^2/\rho \tag{4.69}$$

where P_e is the energy loss (or power loss) per unit volume, K is a geometric constant, B is the maximum induction, f is the frequency, d is the thickness of the narrowest dimension perpendicular to magnetic flux and ρ is the resistivity. A ferromagnetic metal of resistivity $\rho_m \approx 10^{-5}\,\Omega\,m$ has a given amount of eddy-current losses at 50 Hz. If this metal is substituted by a ferrite with resistivity $\rho_f \approx 10^5\,\Omega\,m$ (assuming the same geometry and maximum induction B), it is possible to evaluate the frequency, f_f, at which this ferrite has the same losses as the metal as:

$$f_f = 50(\rho_f/\rho_m)^{1/2} \tag{4.70}$$

leading to $f_f = 5\,MHz$. Assuming that everything else is equal, using a ferrite instead of a metal would permit an increase in the working frequency from 50 Hz to 5 MHz with the same eddy-current losses. Representative values of resistivity for ferrites are shown in Table 4.18.

Eddy currents appear as a consequence of Faraday's law of electromagnetic induction. Consider a cylindrical sample of material, Fig. 4.64. As the current flows in the coil, an ac magnetic field H is applied to the

Table 4.18. *Representative values of electrical resistivity for some ferrites.*

Ferrite	ρ (Ω m) (at 300 K)
Mn–Zn ferrites	10^{-2}–1
Ni–Zn ferrites	10^3–10^8
$BaFe_{12}O_{19}$	$\sim 10^6$
$Y_3Fe_5O_{12}$	$\sim 10^{10}$

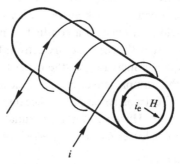

Fig. 4.64. Eddy currents. The variations in current i produced a varying field, H; an emf is generated, resulting in eddy currents, i_e. The intensity of the emf depends on the specimen magnetic permeability. (Adapted from Cullity, 1972.)

material and, depending on its permeability, an ac magnetisation follows the field. As a result of the magnetic flux variations within the sample, a voltage is generated, given by:

$$emf = -d\phi/dt = -A\,dB/dt \tag{4.71}$$

where emf is the voltage, ϕ is the magnetic flux, A the area normal to the flux and B is the induction. This law is valid for any kind of material, regardless of whether it is magnetic or not. In a non-magnetic material (diamagnetic, paramagnetic, antiferromagnetic), flux variations correspond practically to the magnetic field produced by the coil ($B = \mu_0 H$). In a ferro- or ferrimagnetic material, induction is dramatically greater because in these materials permeability can be 10^4 times that of vacuum. The voltage generated is therefore 10^4 times greater. In metals, there are many conduction electrons which can be excited by these voltage variations; in oxides, electron localisation severely limits this process.

Electrical properties

The main conductivity mechanism in ferrites is attributed to electron hopping between Fe^{3+} and Fe^{2+} in octahedral sites. Resistivity in spinels is very sensitive to stoichiometry; a small variation of Fe content in $Zn_{0.7}Ni_{0.3}Fe_{2+\delta}O_{4-\varepsilon}$ (Van Uitert, 1956) results in resistivity variations of $\sim 10^7$, Fig. 4.65. Excess Fe can easily be dissolved in the spinel phase by a partial reduction of Fe from $3Fe_2^{3+}O_3$ to $2Fe^{2+}Fe_2^{3+}O_4$ (and $\frac{1}{2}O_2\uparrow$). A negative δ value maintains all Fe as Fe^{3+} and tends to produce some O vacancies. However, since this last process requires considerable energy, it is very likely that the O vacancy concentration remains quite low and the divalent cation excess appears as a second phase. During sintering of stoichiometric mixtures at high temperatures ($T > 1200\,°C$), there is always the tendency to lose some O_2 with the corresponding reduction in Fe, even in an oxidising atmosphere. In Mn–Zn ferrites, an O_2 atmosphere cannot be used because Mn is oxidised to Mn^{3+} producing cation vacancies and a decrease in resistivity.

In polycrystalline ferrites, grain boundaries can also have an influence on eddy-current losses. In high-permeability Mn–Zn ferrites, for instance, CaO and SiO_2 are used as additives to decrease the Fe^{2+} content and increase the grain boundary resistivity (Berger *et al.*, 1989), both processes limiting eddy-current losses.

The measurement of electrical properties as a function of frequency and their analysis by complex impedance methods (impedance spectroscopy) allow a separation of contributions to impedance from grains, grain boundaries and electrode polarisation (Jonscher, 1983; MacDonald, 1987). This technique therefore permits the separation of the electrical

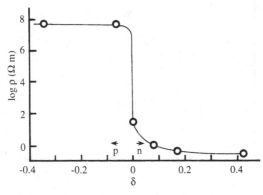

Fig. 4.65. Effect of Fe content, δ, on the electrical conductivity of Ni–Zn ferrites. (Adapted from Van Uitert, 1956.)

properties of grain boundaries from those of grains, Fig. 4.66. In Ni–Zn ferrites, this method has shown that grain boundary resistivity can depend on the thermal history of the sample (Irvine *et al.*, 1990b).

4.8   Magnetooptical properties

When a beam of polarised light is transmitted through, or reflected from, a magnetic material, its polarisation plane is rotated by a certain angle, θ. This phenomenon is useful in a variety of ways; it can be used to observe magnetic structures and to investigate the energy-level structure of magnetic materials. It is also the basis for many technological applications. The interaction can occur by transmission through a magnetic medium, which is known as the *Faraday* effect; or by reflection from a magnetised surface, referred to as the *Kerr* effect.

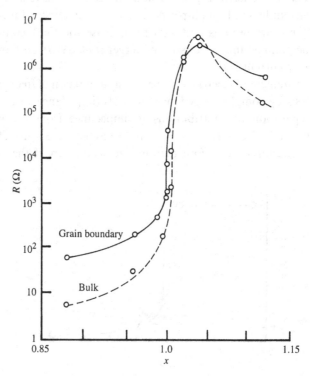

Fig. 4.66. Bulk and grain boundary electrical resistance for $Co_xFe_{3-x}O_4$ ferrites (Ferreira, Furtado & Perdigão, 1992).

Magnetooptical properties

In the Faraday effect, the specimen is usually a thin slab or a thin film placed between two polarisers, Fig. 4.67. The rotation angle is measured by adjusting the polarisers to maximum transmission. Domains appear with different contrasts because the magnetisation has a different orientation in each of them, as in Fig. 4.32. The rotation angle, θ_F, depends on magnetisation and thickness:

$$\theta_F = KMt \tag{4.72}$$

where K is Kundt's constant, M the magnetisation and t the specimen thickness. For each material, K depends on the wavelength of the light and the temperature.

In the Kerr effect, the polarisation rotation can be expressed by a relation similar to Eq. (4.72), except that it is independent of specimen thickness. The Kerr effect can be observed in three geometries: polar, longitudinal and transverse, shown in Fig. 4.68. The rotation angle is largest in the polar geometry, where magnetisation is perpendicular to the specimen surface.

Magnetooptical properties of rare-earth garnets have been extensively studied because these ferrites can be prepared as single-crystal films and

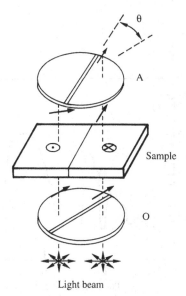

Fig. 4.67. Faraday rotation. A light beam is plane-polarised by polariser O; the plane is rotated by opposite angles, θ, by 180° domains and results in contrast differences when analysed by polariser A.

plates in a wide variety of compositions. In addition to the magnetooptical rotation, the absorption coefficient is also important, since it determines the maximum useful thickness. These properties are strongly dependent on the beam energy, as illustrated for YIG in Fig. 4.69; YIG becomes extremely transparent in the near infrared.

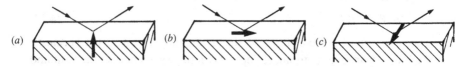

Fig. 4.68. Experimental geometries in the Kerr effect: (*a*) polar; (*b*) longitudinal; and (*c*) transverse arrangements.

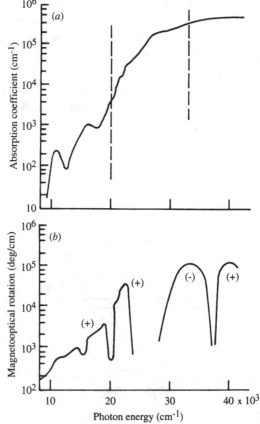

Fig. 4.69. (*a*) Optical absorption and (*b*) magnetooptical rotation coefficients in YIG at 300 K, as a function of photon energy; the broken lines correspond to the visible spectrum (Wemple *et al.*, 1974).

References

Some early experiments on GdIG (Dillon, 1959), near the compensation temperature, showed that it was not the total magnetisation, but sublattice magnetisations which determined the magnetooptical phenomena. A simple model which assumes that magnetooptical spectra can be decomposed into individual contributions from 3d cations in the several sublattices has shown reasonable agreement with experimental results. In YIG, the increase in absorption coefficient at $\sim 20\,000\,\text{cm}^{-1}$ (Fig. 4.69) is attributed to crystal field transitions of Fe^{3+} in both tetrahedral and octahedral sites; the absorption at $33\,000\,\text{cm}^{-1}$ is related to charge transfer from O^{2-} anions towards Fe^{3+} (Dillon, 1982). In magnetite, the contributions from Fe^{3+} in octahedral and tetrahedral sites, and Fe^{2+} in octahedral sites, have been evaluated separately (Nakagawa et al., 1982).

A number of ion substitutions produce a very high Faraday rotation coefficient in garnets: namely, Pr and Nd (Wemple et al., 1974), and Bi and Ce (Gomi, Satoh & Abe, 1989).

References

Aburto, S., Jiménez, M., Marquina, M. L. & Valenzuela, R. (1982). The molecular field approximation in Ni-Zn ferrites. In *Ferrites: Proceedings of the 3rd International Conference on Ferrites, Kyoto, 1980.* Eds H. Watanabe, S. Iida and M. Sugimoto. Center for Academic Publications, Tokyo, pp. 188-91.

Anderson, P. W. (1963). Exchange in insulators: Superexchange, direct exchange and double exchange. In *Magnetism*. Eds G. T. Rado and H. Suhl. Academic Press, New York, pp. 29-83.

Beer, A. & Schwarz, J. (1966). New results on the influence of filter stipulations on the qualities of Mn, Zn ferrites. *IEEE Transactions on Magnetics*, **2**, 470-4..

Berger, M. H., Laval, J. Y., Kools, F. & Roelofsma, J. (1989). Relation between grain boundary structure & hysteresis losses in Mn-Zn ferrites for power applications. In *Advances in Ferrites: Proceedings of the Fifth International Conference on Ferrites, India, 1989*, Vol. 1. Eds C. M. Srivastava and M. J. Patni. Oxford & IBH Publishing Co. PVT Ltd, Bombay, pp. 619-24.

Bethe, H. (1933). Ferromagnetism. *Handbuch der Physik*, **24**, 595-8.

Bickford, L. R., Pappis, J. & Stull, J. L. (1955). Magnetostriction and permeability of magnetite and cobalt-substituted magnetite. *Physical Review*, **99**, 1210-14.

Bitter, F. (1931). On inhomogeneities in the magnetization of ferromagnetic materials. *Physical Review*, **38**, 1903-5.

Bizette, H. & Tsai, B. (1954). Susceptibilités magnétiques principales d'un crystal de side-rose et du fluorure manganeux. *Comptes Rendus de l'Academie des Sciences, Paris*, **238**, 1575-6.

Bloch, F. (1930). On the theory of ferromagnetism. *Zeitschrift Physik*, **61**, 206-19.

Magnetic properties of ferrites

Boucher, B., Buhl, R. & Perrin, M. (1970). Structure magnétique du spinelle antiferromagnétique $ZnFe_2O_4$. *Physica status solidi*, **40**, 171–82.

Broese van Groenou, A., Schulkes, J. A. & Annis, D. A. (1967). Magnetic anisotropy of some nickel zinc ferrite crystals. *Journal of Applied Physics*, **38**, 1133–4.

Brown, W. F. (1984). Tutorial paper on dimensions and units. *IEEE Transactions on Magnetics*, **MAG-20**, 112–17.

Cedillo, E., Ocampo, J., Rivera, V. & Valenzuela, R. (1980). An apparatus for the measurement of initial magnetic permeability as a function of temperature. *Journal of Physics F: Scientific Instruments*, **13**, 383–6.

Christman, J. R. (1988). *Fundamentals of Solid State Physics*. Wiley, New York, p. 357.

Condon, E. U. & Shortley, G. H. (1935). *The Theory of Atomic Spectra*. Cambridge University Press, London.

Cullity, B. D. (1972). *Introduction to Magnetic Materials*. Addison-Wesley, Massachusetts.

Dey, S. & Valenzuela, R. (1984). Magnetic properties of substituted W and X ferrites. *Journal of Applied Physics*, **55**, 2340–2.

Dillon, J. F., Jr (1959). Optical absorptions and rotations in the ferrimagnetic garnets. *Journal de Physique et le Radium*, **20**, 274–7.

Dillon, J. F., Jr (1982). The development of magneto-optical research in garnets and magnetic insulators. In *Ferrites: Proceedings of the Third International Conference on Ferrites, Kyoto, 1980* Eds H. Watanabe, S. Iida and M. Sugimoto. Center for Academic Publications, Tokyo, pp. 743–9.

Dillon, J. F. Jr & Earl, H. E. Jr (1959). Domain wall motion and ferrimagnetic resonance in a manganese ferrite. *Journal of Applied Physics*, **30**, 202–13.

Dionne, G. F. (1970). Molecular field coefficients of substituted yttrium iron garnets. *Journal of Applied Physics*, **41**, 4874–81.

Enz, U. (1958). Relation between disaccommodation and magnetic properties of manganese ferrous ferrite. *Physica*, **24**, 609–24.

Escobar, M. A., Magaña, L. F. & Valenzuela, R. (1983). Analytical prediction of the magnetization curve and the ferromagnetic hysteresis loop. *Journal of Applied Physics*, **54**, 5935–40.

Feenstra, R. M. (1990). *Scanning Tunneling Microscope and Related Methods*, Vol. 184 of NATO ASI Series E: Applied Science. Eds R. J. Behm, N. Garcia and H. Rohrer. Kluwer, North Holland, Amsterdam, pp. 211–40.

Ferreira, A. R., Furtado, C. S. & Perdigão, J. M. (1992). Grain-boundary and grain electrical resistances in $Co_xFe_{3-x}O_4$. *Journal of the American Ceramic Society*, **75**, 1708–11.

Foner, S. (1959). Versatile and sensitive vibrating sample magnetometer. *Review of Scientific Instruments*, **30**, 548–57.

Fuentes, V., Aburto, S. & Valenzuela, R. (1987). Magnetic sublattices in nickel ferrite. *Journal of Magnetism and Magnetic Materials*, **69**, 233–6.

Galt, J. K. (1954). Motion of individual domain walls in nickel–iron ferrite. *Bell System Technical Journal*, **33**, 1023–54.

Gieraltowski, J. (1989). Initial susceptibility frequency spectra and distribution

186

of the grain sizes in Ni–Zn and YIG polycrystalline ferrites. *Journal of Magnetism and Magnetic Materials*, **81**, 103–6.

Gieraltowski, J. & Globus, A. (1977). Domain wall size and magnetic losses in frequency spectra of ferrites and garnets. *IEEE Transactions on Magnetics*, **MAG-13**, 1357–9.

Globus, A. (1962). Influence des dimensions des Parois sur la permeabilité initiale. *Comptes Rendus de l'Academie des Sciences, Paris*, **255**, 1709–11.

Globus, A. (1977). Some physical considerations about the domain wall size theory of magnetization mechanisms. *Journal de Physique (Paris)*, **Cl-38**, Cl-1–15.

Globus, A. & Duplex, P. (1966). Separation of susceptibility mechanisms for ferrites of low anisotropy. *IEEE Transaction on Magnetics*, **2**, 441–5.

Globus, A., Duplex, P. & Guyot, M. (1971). Determination of initial magnetization curve from crystallite size and effective anisotropy field. *IEEE Transactions on Magnetics*, **MAG-7**, 617–22.

Globus, A., Pascard, H. & Cagan, V. (1977). Distance between ions and fundamental properties in ferrites. *Journal de Physique*, **Cl-38**, Cl-163–8.

Gomi, M., Satoh, K. & Abe, M. (1989). New garnet films with giant Faraday rotation. In *Advances in Ceramics: Proceedings of the Fifth International Conference on Ferrites, India*, Vol. 2. Eds C. M. Srivastava and M. J. Patni. Oxford & IBH Publishing Co. PVT Ltd, Bombay, pp. 919–24.

Gorter, E. W. (1954). Saturation magnetization and crystal chemistry of ferrimagnetic oxides. *Philips Research Reports*, **9**, 295–320.

Guillaud, C. & Sage, M. (1951). Propriétés magnétiques des ferrites mixtes de magnésium et de zinc. *Comptes Rendus de l'Academie des Sciences, Paris*, **232**, 944–6.

Guyot, M. & Globus, A. (1977). Determination of the domain wall energy and the exchange constant from hysteresis in ferrimagnetic polycrystals. *Journal de Physique (Paris)*, **Cl-38**, Cl-157–62.

Herring, C. (1962). Critique of the Heitler–London method of calculating spin couplings at large distances. *Reviews of Modern Physics*, **34**, 631–45.

Hund, F. (1927). *Linienspectren und Periodische System der Elemente*, Springer, Berlin.

Hwang, C., Laughlin, D. E., Mitchell, P. V., Layadi, A., Mountfield, K., Snyder, J. E. & Artman, J. O. (1986). TME investigation of cobalt–chromium film microstructure. *Journal of Magnetism and Magnetic Materials*, **54–57**, 1676–8.

Irvine, J. T. S., West, A. R., Amano, E., Huanosta, A. & Valenzuela, R. (1990a). Characterisation of magnetic materials by impedance spectroscopy. *Solid State Ionics*, **40/41**, 220–3.

Irvine, J. T. S., Huanosta, A., Valenzuela, R. & West, A. R. (1990b). Electrical properties of polycrystalline nickel zinc ferrites. *Journal of the American Ceramic Society*, **73**, 729–32.

Jonscher, A. K. (1983). *Dielectric Relaxation in Solids*, Chelsea Dielectric Press, London.

Kanamori, J. (1963). Anisotropy and magnetostriction of ferromagnetic and antiferromagnetic materials. In *Magnetism*, Vol. 1. Eds G. T. Rado and H. Suhl. Academic Press, New York, pp. 127–203.

Kaya, S. (1928). On the magnetization of single crystals of cobalt. *Scientific Reports, Tohoku University*, **17**, 1175–7.

Kittel, C. (1986). *Introduction to Solid State Physics*, 6th edition. Wiley, New York.

Kittel, C. & Galt, J. K. (1956). Ferromagnetic domain theory. *Solid State Physics*, **3**, 437–564.

Koike, K. & Hayakawa, K. (1985). Domain observation with spin-polarized secondary electrons. *Journal of Applied Physics*, **57**, 4244–8.

Landau, L. & Lifshitz, E. (1935). On the theory of the dispersion of magnetic permeability in ferromagnetic bodies. *Physikalische Zeitschrift der Sowjetunion*, **8**, 153–69.

Lilot, A. P., Gérard, A. & Grandjean, F. (1982). Analysis of the superexchange interaction paths in the W-hexagonal ferrites. *IEEE Transactions on Magnetics*, **MAG-18**, 1463–5.

Lin, I-N., Mishra, R. K. & Thomas, G. (1984). Interaction of domain walls with microstructural features in spinel ferrites. *IEEE Transactions on Magnetics*, **MAG-20**, 134–9.

Lotgering, F. K., Vromans, P. H. G. M. & Huyberts, M. A. H. (1980). Permanent-magnet material obtained by sintering the hexagonal ferrite $W = BaFe_{18}O_{27}$. *Journal of Applied Physics*, **51**, 5913–18.

Luborsky, F. E. (1961). Development of elongated particle magnets. *Journal of Applied Physics*, **32**, 171S–83S.

MacDonald, J. R. (1987). *Impedance Spectroscopy*. Wiley, New York.

Magaña, L. F., Escobar, M. A. & Valenzuela, R. (1986). Effect of the grain size distribution on the ferromagnetic hysteresis loops. *Physica status solidi (a)*, **97**, 495–500.

Mee, C. D. & Jeschke, J. C. (1963). Single-domain properties in hexagonal ferrites. *Journal of Applied Physics*, **34**, 1271–2.

Moulson, A. J. & Herbert, J. M. (1990). *Electroceramics*. Chapman and Hall, London, p. 413.

Nakagawa, Y., Hori, H., Goto, T. & Miyata, N. (1982). Magneto-optical spectra of zinc-, manganese- and nickel-substituted magnetite. In *Ferrites: Proceedings of the Third International Conference on Ferrites, Kyoto, 1980*. Eds H. Watanabe, S. Iida and M. Sugimoto. Center for Academic Publications, Tokyo, pp. 110–14.

Néel, L. (1932). Influence des fluctuations du champ moléculaire sur les proprietés magnétiques de corps. *Annales de Physique*, **18**, 5–105.

Néel, L. (1948). Proprietés magnétiques des ferrites: ferrimagnétisme et antiferromagnétisme. *Annales de Physique*, **3**, 137–98.

Ohta, K. (1963). Magnetocrystalline anisotropy and initial permeability of Mn–Zn–Fe ferrites. *Journal of the Physical Society of Japan*, **18**, 685–90.

Pankhurst, Q. A., Jones, D. H., Morrish, A. H., Zhou, X. Z. & Corradi, A. R. (1989). Cation distribution in Co–Ti substituted barium ferrite. In *Advances in Ferrites: Proceedings of the Fifth International Conference on Ferrites, India, 1989*, Vol. 1. Eds C. M. Srivastava and M. J. Patni. Oxford & IBH Publishing Co. PVT Ltd, Bombay, pp. 323–7.

References

Pauthenet, R. (1952). Aimantation spontanée de ferrites. *Annales de Physique*, 7, 710–45.

Polcarova, M. (1969). Application of X-ray diffraction topography to the study of magnetic domains. *IEEE Transactions on Magnetics*, **MAG-5**, 536–44.

Rao, L. M. (1989). Neutron diffraction survey of perturbed magnetic ordering in disordered spinels. In *Advances in Ferrites: Proceedings of the Fifth International Conference on Ferrites*, India, 1989. Eds C. M. Srivastava and M. J. Patni. Oxford & IBH Publishing Co. PVT Ltd, Bombay, pp. 385–97.

Rave, W., Schafer, R. & Hubert, A. (1987). Quantitative observation of magnetic domains with the magneto-optical Kerr effect. *Journal of Magnetism and Magnetic Materials*, **65**, 7–14.

Roess, E. (1971). Magnetic properties and microstructure of high permeability Mn–Zn ferrites. In *Ferrites: Proceedings of the International Conference, Japan, July 1970*. Eds Y. Hoshino, S. Iida and M. Sugimoto, University of Tokyo Press, Tokyo, pp. 203–9.

Röschmann, P. & Hansen, P. (1981). Molecular field coefficients and cation distribution of substituted yttrium iron garnets. *Journal of Applied Physics*, **52**, 6256–69.

Sáenz, J. J., García, N., Grütter, P., Meyer, E., Heinzelmann, H., Wiesendanger, R., Rosenthaler, L., Hidber, H. R. & Güntherodt, H. J. (1987). Observation of magnetic forces by the atomic force microscope. *Journal of Applied Physics*, **62**, 4293–5.

Satya Murthy, N. S., Natera, M. G., Youssef, S. I. & Begum, R. J. (1969). Yafet–Kittel angles in zinc–nickel ferrites. *Physical Review*, **181**, 969–77.

Semat, H. & Albright, J. R. (1972). *Introduction to Atomic and Nuclear Physics*, 5th edition. Chapman and Hall, New York, p. 250.

Serres, A. (1932). Recherches sur les moments atomiques. *Annales de Physique*, **17**, 5–95.

Shull, C. G., Wollan, E. O. & Koehler, W. C. (1951). Neutron scattering and polarization by ferromagnetic materials. *Physical Review*, **84**, 912–21.

Smart, J. S. (1966). *Effective Field Theories of Magnetism*. Saunders, Philadelphia.

Smit, J. & Wijn, H. P. J. (1961). *Les Ferrites*. Bibliothéque Technique Philips, Dunod, Paris.

Srivastava, C. M., Shringi, S. N., Srivastava, R. G. & Nanadikar, N. G. (1976). Magnetic ordering and domain wall relaxation in zinc-ferrous ferrites. *Physical Review B*, **14**, 2032–40.

Stadnik, Z. M. & Zarek, W. (1979). On the exchange integrals in nickel ferrite. *Physica status solidi (b)*, **91**, k83–5.

Stoner, E. C. & Wohlfarth, E. P. (1948). A mechanism of magnetic hysteresis in heterogenous alloys. *Philosophic Transactions of the Royal Society, London*, **A240**, 599–642.

Tauber, A., Megill, J. S. & Shappiro, J. R. (1970). Magnetic properties of $Ba_2Zn_2Fe_{28}O_{46}$ and $Ba_2Co_2Fe_{28}O_{46}$ single crystals. *Journal of Applied Physics*, **41**, 1353–4.

Turilli, G., Licci, F., Rinaldi, S. & Deriu, A. (1986). Mn^{2+}, Ti^{4+} substituted barium ferrite. *Journal of Magnetism and Magnetic Materials*, **59**, 127–31.

189

Valenzuela, R. (1980). A sensitive method for the determination of the Curie temperature in ferrimagnets. *Journal of Materials Science*, **15**, 3173–4.

Van Loef, W. & Broese van Groenou, A. (1965). On the sublattice magnetization of $BaFe_{12}O_{19}$. In *Proceedings of the International Conference on Magnetism, Nottingham, 1964*. Institute of Physics and The Physical Society, London, pp. 646–9.

Van Uitert, L. G. (1956). Dielectric properties and conductivity of ferrites. *Proceedings of the Institute of Radio Engineers*, **44**, 1294–303.

Vella-Coleiro, G. P., Smith, D. H. & Van Uitert, L. G. (1972). Resonant motion of domain walls in yttrium gadolinium iron garnets. *Journal of Applied Physics*, **43**, 2428–30.

Wemple, S. H., Blank, S. L., Seman, J. A. & Bielsi, W. A. (1974). Optical properties of epitaxial iron garnet thin films. *Physical Review B*, **9**, 2134–44.

Wiesendanger, R., Shvets, I. V., Bürgler, D., Tarrach, G., Güntherodt, H. J., Coey, J. M. D. & Gräeser, S. (1992). Topographic and magnetic-sensitive scanning tunneling microscope study of magnetite. *Science*, **255**, 583–6.

Williams, H. J., Bozorth, R. M. & Shockley, W. (1949). Magnetic domain patterns on single crystals of silicon iron. *Physical Review*, **75**, 155–78.

Williams, H. J., Shockley, W. & Kittel, C. (1950). Studies on the propagation velocity of a ferromagnetic domain boundary. *Physical Review*, **80**, 1090–4.

Yafet, Y. & Kittel, C. (1952). Antiferromagnetic arrangements in ferrites. *Physical Review*, **87**, 290–4.

Zijlstra, H. (1967). *Experimental Methods in Magnetism*. North-Holland, Amsterdam. Vol. 2.

5 Applications of ferrites

Magnetic ceramics represent an important fraction of the magnetic industry; in the US, an estimated 40% of the total hard magnetic materials market value is dominated by ferrites, and in spite of the continuous development of new materials, ferrite consumption is still growing. In soft material applications, ferrites participate with an estimated 20% of the market value. In 1990, the estimated world production was 159 500 metric tons of soft ferrites, and 431 100 metric tons of hard ferrites (Ruthner, 1989). In addition to the versatility of ferrites, there are two essential factors which explain this success: the low electrical conductivity, and the low production cost. The market value of ferrites ($\sim$\$3/kg) is very low compared with other electroceramics: \$33/kg for varistors, \$330/kg for thermistors and \$1100/kg for ceramic capacitors (Cantagrel, 1986).

5.1 Permanent magnet devices

The significant properties for permanent magnet materials are the remanent induction, B_r, and the coercive field, H_c; the aim is to provide a high magnetic flux in a given volume (B_r), and a high resistance to change when subjected to strong fields (H_c). A high magnetic flux can also be produced by means of electromagnetic devices, with intensities larger than those of hard magnets; however, in many applications it is more economic and convenient to use a permanent magnet. It provides a constant field with no energy consumption; there is an additional energy saving from ohmic losses. The leading ceramic materials for permanent magnetic applications are Ba and Sr hexaferrites.

The largest single application of anisotropic hexaferrites is in *loudspeakers*, Fig. 5.1. The ferrite is an anisotropic ring, axially magnetised. The signal to be transformed to sound, typically from an amplifier, flows through the coil fixed to the end of the speaker cone; the interaction between the magnetic flux produced by the ferrite element and the current results in an axial force on the speaker cone, which vibrates according to the electrical signal. The ferrite ring acts therefore as a spring force on the cone coil.

The movement in an electric motor is produced by the torque of a magnetic field on the rotor. The magnetic field can be produced by a current flowing through a coil; however, in small dc motors, it is more convenient to use a permanent magnet to drive the rotor, resulting in a smaller and cheaper device. There are many applications for dc, or *permanent-magnet motors*; an expensive car, for instance, can have some 20 of them (windshield wipers, windows, fans, etc.). The permanent magnet element is a curved segment, Fig. 5.2, with a magnetisation normal to the

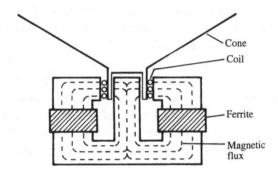

Fig. 5.1. Ferrite permanent magnet in a loudspeaker.

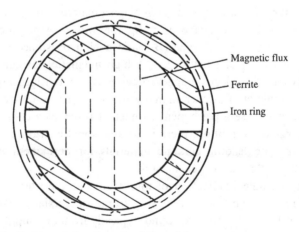

Fig. 5.2. Ferrite segments in a dc motor (schematic).

192

segment planes. A soft Fe ring provides a mechanical structure, as well as a convenient medium for completing the magnetic circuit. Since demagnetising fields in motors are stronger than in loudspeakers, Sr ferrite is preferred to Ba ferrite for these applications.

In some specialised applications such as floppy-disc drivers, printers and other computer peripherals, a *stepping motor* is needed. These motors rotate a fraction of a revolution upon application of an electric pulse; an accurate step angle under a high output torque is required. This is achieved with a radially magnetised rotor. To produce such a rotor, a rolling technology has been used, Fig. 5.3 (Torii, Kobayashi & Okuda, 1982). The hexaferrite particles were single-domain platelets with the *c*-axis (and therefore the easy axis) normal to the plane; after mixing with an organic binder, a rolling operation was performed. During this operation, which was repeated 5–10 times to obtain optimal results, the particles were subjected to a strong shearing action resulting in a preferential orientation. The tape obtained was wound on a mandrel at 50–70 °C under pressure; the binder was reduced at 300 °C before sintering. A multipole magnetisation process was used to produce the final magnetic configuration. A similar technology can be used for other applications such as a refrigerator gasket; in this case, the rolling medium is a rubber which leads to the final product by hardening of the tape.

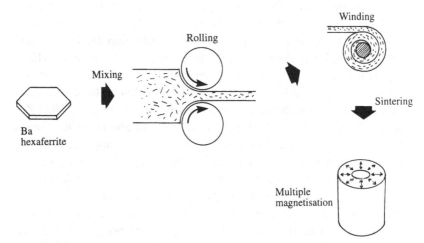

Fig. 5.3. Rolling technology to produce radially magnetised rotors for stepping motors (Torii, Kobayashi & Okuda, 1982).

5.2 Linear response applications

This group of soft-ferrite applications is based on their ability to transform ac signals of small amplitude into substantially large variations of magnetic flux. The fact that at low fields the initial permeability is a linear function of the field explains the name; these devices are also known as small-signal applications. There are other materials, such as metallic alloys (see Chapter 6), which possess permeability values considerably higher than the typical values of ferrites; however, as frequency increases, conductivity losses prevent efficient use of metallic materials.

In all these applications, a high permeability value in the room-temperature range (-10–$50\,^\circ\text{C}$) is required. If a typical temperature dependence of the initial permeability is considered, such as the one shown in Fig. 4.48, it appears that the high μ_i range is just below the Curie temperature. At first sight it seems that, in Ni–Zn ferrites, for instance, it would be enough to prepare a composition rich in Zn to obtain high μ_i values at room temperature (see Fig. 4.46). However, in this temperature range there are also large permeability variations, which make such materials unsuitable for this application. A simple solution is to incorporate some additives in the ferrite which result in a round shape at the Curie transition, almost always with an associated decrease in permeability. This phenomenon occurs presumably by producing composition gradients within the grains, in such a way that there is no longer a unique Curie temperature, but a distribution of T_C, Fig. 5.4. Instead of a single $\mu_i(T)$ curve, the combination of a family of curves is obtained.

There exists another technique for controlling the temperature coefficient of the permeability in the working temperature range, based on the effects of a positive anisotropy cation. As discussed in Section 4.4.1, some cations, particularly Co^{2+} and Fe^{2+}, have a positive contribution to the magneto-crystalline anisotropy. Since anisotropy is negative in almost all the cubic ferrites, the presence of these cations in concentrations higher than a certain limit results in a change of sign of the anisotropy constant, and therefore in a secondary maximum in permeability, Fig. 4.49. By varying the amount of such a cation, it is possible to obtain permeability curves with the desired temperature variation in the working range, Fig. 5.5. Since many other materials show a positive temperature coefficient (i.e., an increase in the magnitude of a given property as a function of temperature), a negative temperature coefficient can be useful in compensating for the behaviour of the other components.

Linear response applications

The requirements of high permeability and high working frequency lead to a compromise, since for a given composition, an increase in grain size leads simultaneously to an increase in permeability and a decrease in domain wall relaxation frequency. The choice of the basic composition

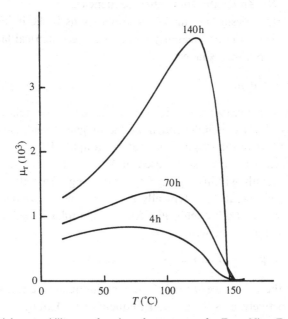

Fig. 5.4. Initial permeability as a function of temperature for $Zn_{0.64}Ni_{0.36}Fe_2O_4$ ferrites. Short sintering times (at 1150 °C) result in a lower degree of homogeneity and smaller permeability variations near the Curie point (Globus & Valenzuela, 1975).

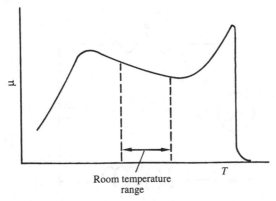

Fig. 5.5. Negative temperature coefficient of initial permeability by adding a positive-anisotropy cation (schematic).

also represents a compromise, since both wall relaxation frequency and permeability are related to magnetocrystalline anisotropy. Relaxation frequency increases with anisotropy, and permeability decreases with K_1. The general tendency is to use Mn–Zn ferrites for frequencies up to 1 MHz, and Ni–Zn ferrites for higher frequencies.

The important design parameter in applications is the inductance, L, which depends on the permeability through the geometrical factor, k, of the particular specimen shape:

$$L = kn^2\mu_0\mu_i \tag{5.1}$$

where n is the number of coil turns and μ_i is the relative, initial permeability. Since μ_i is independent of the magnetic field, L is constant and independent of the magnetising current amplitude.

Ferrite devices are extensively used in *filter circuits*, where the aim is to separate signals within a given frequency range from the rest of the spectrum. This operation is typically performed with a *resonant circuit*, which is simply an *LCR* circuit, Fig. 5.6. The complex impedance, Z^*, of this arrangement can be written:

$$Z^* = R + j(\omega L - 1/\omega C) \tag{5.2}$$

where R is the resistance, L and C are the inductance and capacitance values, respectively, ω is the angular frequency $(2\pi f)$ and j is $(-1)^{1/2}$. At the resonance frequency, ω_s,

$$\omega_s L = 1/\omega_s C \tag{5.3}$$

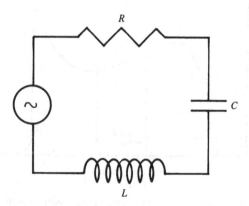

Fig. 5.6. A resonant, *LCR* circuit.

Z^* is a minimum and the current flowing through the circuit becomes a maximum. Signals of frequency different to ω_s are strongly attenuated, and the circuit becomes a frequency filter.

The ferrite inductors are shaped as *pot cores*, Fig. 5.7, formed of two matching halves. The central part is shorter than the outer surfaces; when they are in contact, a gap appears in the centre of the device. This central part has a hole, such that a magnetic rod can be inserted into it, in an adjustable position. An adjustable gap is quite useful in pot cores; in addition to allowing a fine calibration of the final inductance value of the device, it decreases its temperature sensitivity and increases the working frequency. The presence of a gap leads to a lower permeability value because a demagnetising field is generated, decreasing the value of the total applied field:

$$H_T = H - NM \tag{5.4}$$

where H is the applied field, H_T is the total field, N is the demagnetising factor and M is the magnetisation. The presence of a gap leads to a decrease in permeability; however, since the demagnetising field is proportional to the magnetisation, permeability variations (produced by temperature changes, for instance) are attenuated. Another beneficial effect of the gap is that the relaxation frequency increases. This behaviour of the permeability can be explained in terms of the losses. For some practical purposes, it is useful to define the energy losses as:

$$\tan \delta = \frac{\text{energy dissipated}}{\text{energy stored}} = \frac{\mu''}{\mu'} \tag{5.5}$$

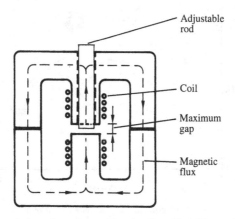

Fig. 5.7. Cross-sectional view of a pot core, showing the adjustable gap (schematic).

where the loss factor is tan δ, and μ'' and μ' are the imaginary and the real parts of the complex permeability, respectively. In high-permeability materials ($\mu > 100$), a general trend is that the factor (tan δ/μ_0) $\approx$ constant, where μ_0 is the low frequency value of permeability. A decrease in permeability as a result of a gap leads not only to a decrease in permeability, but also to a decrease in the loss factor, Fig. 5.8. The frequency range of a ferrite is therefore extended at the expense of its permeability.

Ferrites are also used in *antennas*, to transform an electromagnetic signal, transmitted through the air, into an electric signal. Antennas can be constructed with other types of materials; however, ferrites provide a compact device well adapted to small radio receivers. Ferrite antennas are usually a simple rod with a wound coil. As in almost all the ac applications, the resistivity of the ferrite is critical to prevent eddy-current losses.

An important phenomenon which can seriously affect the performance of a ferrite device is the *dimensional resonance*, which occurs when the wavelength of the signal propagating through the ferrite is comparable to the dimensions of the specimen. The propagation velocity of

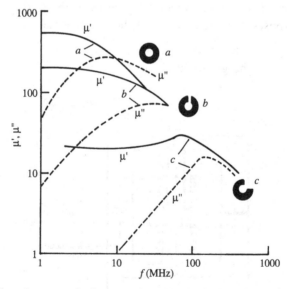

Fig. 5.8. Effect of a gap on the frequency dependence of the real and imaginary permeabilities of a $Zn_{0.64}Ni_{0.36}Fe_2O_4$ ferrite (Verweel, 1971).

electromagnetic waves, v, in any medium can be written as:

$$v = (\mu\varepsilon)^{-1/2} \qquad (5.6)$$

where μ is the permeability and ε the permittivity of the medium. The propagation velocity can also be expressed in terms of the frequency, f, and the wavelength, λ, of the signal:

$$v = f\lambda \qquad (5.7)$$

As an example, in a material with $\mu_r = 100$ and $\varepsilon_r = 1 \times 10^4$ (absolute $\mu = 4\pi \times 10^{-10}$ H/m and $\varepsilon = 8.854 \times 10^{-8}$ F/m, respectively), the propagation velocity is $\approx 3 \times 10^5$ m/s. When a signal of $f = 15$ MHz propagates through this medium, dimensional resonance can be established if the solid has a dimension comparable to half the wavelength, $\lambda/2 = 3 \times 10^5$ (m/s)/$(2 \times 15 \times 10^6$ (1/s)) $= 1$ cm. At this resonance condition the apparent permeability vanishes. To avoid dimensional resonance, antenna rods are designed with a ribbed cross-section, which limits the resonance conditions to a small fraction of the rod.

Ferrite thin films can be included in integrated circuits where an inductor is required. A metallic coil is deposited on an epitaxial YIG film by sputtering, with a 'meander' pattern. To increase the effective inductance value, a second ferrite film is placed on top of the coil, in a 'sandwich' configuration (Arai *et al.*, 1991). Plating techniques (see Section 3.6.3) have also been used to deposit ferrite thin films on GaAs integrated circuits (Abe *et al.*, 1988).

5.3   *Power applications*

A particular application requiring soft ferrites that has rapidly grown in importance in the last few years is power supplies for computers, peripherals and small instruments. A compact and efficient power unit can be obtained by using a technique known as *switched-mode* power supply (SMPS). It may seem peculiar that in this technique, involving a dc to dc conversion, one of the key elements is a high-frequency transformer.

In the switched-mode technique, the mains power signal is rectified and smoothed; it is then switched as rectangular pulses at a frequency of ~ 25 kHz into a ferrite transformer to obtain the desired voltage; finally, it is rectified and smoothed to provide the required power to the instrument. The combination of modern transistor switching circuits with power transformer leads to substantial reductions in the size and weight of the device, with a high efficiency. SMPSs providing 2 kW are currently available. The current trend is, however, to increase further both the efficiency and the power output, which points to an increase in the working frequency. The power output is approximately proportional to the frequency (Snelling, 1988).

Some of the problems related to the switching step, such as switching losses and generation of parasitic harmonics, can be solved by using an *LC* resonant circuit (Snelling, 1989). The requirements for the ferrite element are a high saturation flux density, a small coercive field (low hysteresis loss), low magnetostriction, high resistivity and a Curie temperature in excess of 150 °C (the limiting working temperature is 100 °C). Mn–Zn ferrites with a number of additives have been investigated, since their coercive field is lower than that of Ni–Zn ferrites; promising results have been obtained with $Ti^{4+} + Co^{2+}$ substitution (Stijntjes, 1989), $Sn^{4+} + Ti^{4+}$ substitution (Hanke & Neusser, 1984), and Ca^{2+} additives, in combination with HIP pressing (Sano, Morita & Matsukawa, 1989). The ferrite shapes form a closed magnetic circuit; however, to allow a fast winding, they are usually formed of two pieces: 'C' or 'E' shapes with a matching bar, Fig. 5.9. Matching surfaces are carefully ground to prevent any gap between the two pieces, since this application requires the highest magnetic flux density.

TV deflection yokes are used in display monitors to focus electrons through the interaction with a strong magnetic field, and create an image on the screen. The requirements for the ferrite element in yokes are similar to those for high-frequency transformers since the image resolution has a direct relation with the frequency needed in the horizontal scanning. A resolution of 2.5×10^6 pixels is achieved with a horizontal frequency of

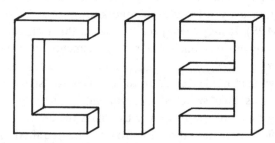

Fig. 5.9. Ferrite shapes for high-frequency core transformers.

90 kHz; doubling that resolution requires a scanning frequency of 130 kHz. Ni–Zn ferrites with Cu^{2+}, Mg^{2+} and Ti^{4+} additives have been investigated for high-frequency deflection yokes (Kobayashi *et al.*, 1992). Besides the use of additives, a small Fe^{3+} deficit with respect to the stoichiometric composition substantially reduces the conductivity losses, presumably by keeping all the Fe in a trivalent state.

5.4 *Magnetic recording*

5.4.1 *Core memories and bubbles*

The basis of information storage technology is the hysteresis loop. The 'memory' is related to the remanent state of a ferromagnetic or ferrimagnetic material, since it is directly related to the magnitude and direction of the most recent applied field. In the early 1970s, information storage technology was dominated by *ferrite core memories*, possessing a 'square' hysteresis loop, Fig. 5.10. In square-loop ferrites, the coercive field is well defined; the remanent state is inverted only if an opposing field $H \geq H_m$ is applied (H_m is slightly greater than H_c, since the slope in the inversion process is never vertical). There are therefore two distinct remanent states, $+B_r$ and $-B_r$, which in a digital device can be taken as '0' and '1', respectively.

Ferrite core memories consist of a large matrix of ferrite rings with two

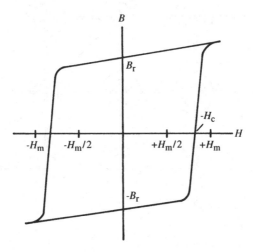

Fig. 5.10. 'Square' hysteresis loop.

wires, one set in the X-direction, and the other in the Y-direction, Fig. 5.11. To set all the cores to '0' ($+B_r$), a current strong enough to apply a field $+H_m$ on the cores is produced in the X wires. Each ferrite ring is thus subjected to a field $+H_m$ and regardless of its previous state, adopts a $+B_r$ state when the field is removed. Instead of a dc current, it is more convenient to apply a current pulse. To write a '1' in the $X_5 Y_8$ ring, for instance, a pulse $-H_m/2$ is sent simultaneously in the fifth column and the eighth row. All the ferrite rings in the fifth column, and those in the eighth row experience a field of $-H_m/2$, which does not affect their remanent state; once the pulse is removed, all these ferrites recover their previous $+B_r$ state. However, the ferrite ring in the $X_5 Y_8$ position was effectively subjected to a field of $2(-H_m/2)$, and therefore changes its remanent state to $-B_r$ ('1'). To read the memory, pulses of $+H_m$ are sent to each ring and a third wire senses if a change of flux occurs, which indicates the state of that particular ring before the pulse application. Matrices of 10 000 cores, with capacities of $\sim 10^6$ bits, have been produced. Spinels in the $(\mathrm{Mn}–\mathrm{Zn}–\mathrm{Mg})\mathrm{Fe}_2\mathrm{O}_4$ system show the desired properties for this application.

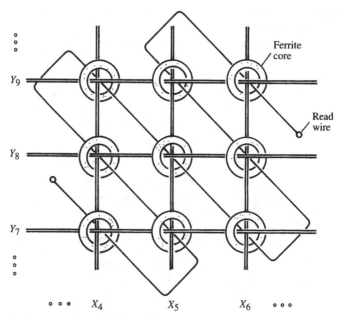

Fig. 5.11. Ferrite core memory (schematic).

Magnetic recording

An advantage of these ferrite memories is that they are 'non-volatile'; the stored information is maintained without the need for an external energy source, which is not the case for the semiconductor memories currently used in computers. However, semiconductor memories possess higher storage densities, shorter access times and lower error rates, and have displaced ferrite memories from the market.

Another application of ferrites, although far less commercially successful than initially expected, is *bubble memories*. In epitaxially grown thin film garnets, the mismatch between the magnetic film and the non-magnetic substrate generates stress and leads to a uniaxial anisotropy; the resulting easy direction is perpendicular to the film plane. The application of a strong perpendicular field saturates the film; slow removal of the field leads to the formation of some small, cylindrical domains with magnetisation antiparallel to the saturated regions, Fig. 5.12. An important effect is that there is a repulsive interaction between these small domains, preventing their coalescence into a single, larger domain.

These cylindrical domains, which appear as bubbles when observed in an optical microscope, are easily displaced by an additional field, *h*, and a system of 'T'-shaped metallic patterns (deposited by evaporation of a ferromagnetic alloy on the film surface), Fig. 5.13. Bubbles are attracted to the 'T' side with opposite polarisation; since *h* is usually a rotating field, the bubbles move around the metallic pattern. In a different region of the film, a strong, localised field can create or destroy bubbles, depending on its orientation. The bubble stream is thus constantly moving, and the information is stored as spaces ('1') between bubbles ('0'). The reading of the memory is performed by means of a metal whose electrical resistance depends on the field; the moving chain of bubbles modulates the current flowing through this metal and therefore provides the information stored.

Semiconductor memories, however, have some clear advantages over bubble systems: they are cheaper, their access time is shorter and there is no need for magnetic fields to maintain the memory elements. Bubble memories are not a serious competitor with semiconductor memories.

5.4.2 Magnetic recording processes

The current magnetic recording technology is based on tapes and discs; the 'writing' and 'reading' processes are relatively simple to understand. They are essentially the same in audio, video and computer recording;

they need a write/read magnetic *head*, which is a small electromagnet, and a recording *medium* (typically a disc or a tape), where the information is stored and/or read. The head has a small gap, typically 0.3 µm, perpendicular to the tape plane, Fig. 5.14. In the writing process, the electrical current representing the information to be stored induces a magnetic field in the head, which appears in the gap. Some of the flux, known as the *fringing* field, completes the magnetic circuit outside the gap; it is this fringing field which produces magnetisation changes in the tape as it moves in front of the head.

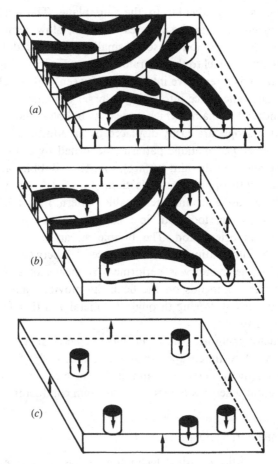

Fig. 5.12. Epitaxial film for bubble memory: (*a*) no applied field; (*b*) small field (oriented antiparallel to the black domains); (*c*) by saturating and decreasing the field, cyclindrical (bubble) domains are formed.

Magnetic recording

Tapes and discs are made of single-domain particles deposited on a plastic substrate, with their easy direction in the tape or disc plane, Fig. 5.15. The fringing field variations from the head result in realignment of the magnetisation of the particles; they form a record of the direction and magnitude of the fringing field, and therefore of the original signal.

In the reading process, the information stored in the tape in the form of particular magnetisation configurations leads to variations in the flux density of the head, which are transformed into voltage variations in the coil. The signal is amplified and depending on the particular device, activates a speaker (audio), a TV system (video) or a processing unit (a computer). In *analogue* recording, the aim is accurately to translate the signal amplitude variations into magnetisation configurations with the same amplitude (in the length of the tape); in reading, the aim is to recreate a signal from the magnetisation configurations in the tape, as similar to the original signal as possible. In *digital* recording, Fig. 5.16, the amplitude of the magnetisation configurations written on the disc is not critical; the important feature is the sequence in the magnetisation sign.

The processes briefly described above are known as *parallel* recording, since the disc (or tape) particle magnetisation is parallel to its plane; a

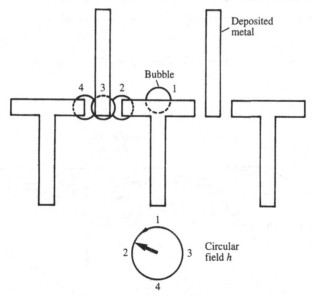

Fig. 5.13. Bubble movement. The successive positions of a bubble are indicated for the corresponding orientations of the field. (Adapted from Moulson & Herbert, 1990.)

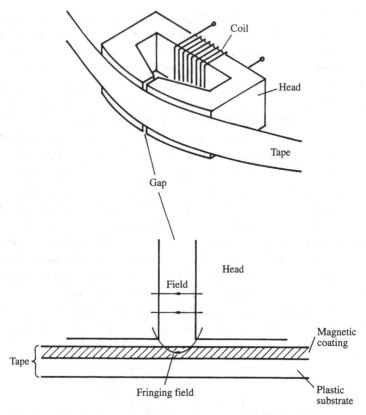

Fig. 5.14. Recording head and tape, showing the gap and the fringing field.

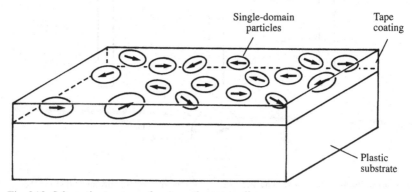

Fig. 5.15. Schematic structure of a magnetic tape or disc.

different technology known as *perpendicular* recording has been attracting attention since the early 1980s, because it can lead to considerably higher storage densities, magnetisation variations and resolution. The basis of this technology is to achieve a recording medium with the easy magnetisation direction perpendicular to the substrate plane, Fig. 5.17. The written 'bits' become denser and sharper since they coincide directly with magnetisation variations; in parallel recording, this correlation is weaker because it is established through the fringing field.

5.4.3   Magnetic recording heads

The optimal properties for a magnetic recording head are summarised in Table 5.1. A low coercive field is related to a high permeability, but does not necessarily imply a high saturation magnetisation. In addition to a high permeability (i.e a soft, easily magnetisable material), it is desirable to induce a strong magnetic flux in the head gap, in order to have a distinct imprint of the information on the recording medium. A low remanence is required to prevent any writing on the tape when the coil signal is zero. A wide frequency spectrum is needed since signal frequencies can be as high as 10 MHz. An additional requirement for analogue recording heads

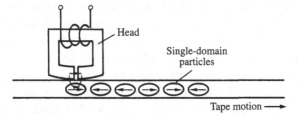

Fig. 5.16. Digital recording (schematic). (Adapted from Cullity, 1972.)

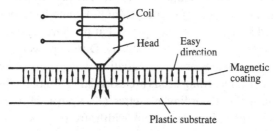

Fig. 5.17. Perpendicular recording mode (schematic).

Table 5.1. *Optimal properties of*
magnetic recording heads.

High permeability
High saturation magnetisation
Low coercive field
Low remanence
High relaxation frequency
Analogue recording: linear response
Chemical stability
High wear resistance

is the linearity of response to avoid any distortion in both the writing and reading processes. The wear and chemical stability characteristics of a head are important, since they define its useful life. The wear arises because the tape is effectively in contact with the head. Audio and video tape recorder (VTR) heads even have a curvature to increase the physical contact with the tape. The relative hardnesses of tape and head are therefore important.

Magnetic ceramics can satisfy virtually all the required characteristics for a recording head; they show a distinct advantage over metallic materials with regard to stability and wear properties. Since ceramics have a combination of covalent and ionic bonding, they are extremely hard and therefore, resistant to wear. They are oxides and are therefore chemically stable, since no further oxidation can be expected.

Polycrystalline, HIP-pressed Ni–Zn ferrites were used in the first recording heads, especially in audio and VTR recorders; they were replaced by Mn–Zn because of their lower coercive field. A hot-pressing process has been used to manufacture high-density Mn–Zn ferrites with a strong grain orientation (Kugimiya, 1990). The raw materials used were small, thin platelets of α-Fe_2O_3 and γ-MnOOH, and ZnO powder. They were wet-mixed and uniaxially hot-pressed at pressures and temperatures about 500 kg/cm^2 and 1400 °C, respectively. During pressing, the platelets of both Fe and Mn oxides are preferentially oriented perpendicular to the pressing axis. These recording heads are essential in doubling the recording time of initial 'Beta' recording cassettes (Kugimiya, Hirota & Bando, 1974). However, the reduction in track widths from ~ 100 to ~ 10 μm led to the use of ferrite single crystals in recording heads; further reductions in track width (and therefore in head gap) down to $\sim 0.3\ \mu$m are expected in the near future (Kugimiya, 1990).

Hard-disc drive heads are subjected to a heavier use and therefore need a different design in order to prevent wear and extend the useful life of both the disc and the head. Hard-disc drive heads (also known as Winchester heads) make no physical contact with the hard disc. They are designed to fly very close to the disc, supported by the air flow resulting from the disc rotation, Fig. 5.18. The magnetic element in Winchester heads is usually a Mn–Zn ferrite.

5.4.4   Magnetic recording media

Magnetic recording tapes and discs are usually made of small, single-domain particles. A high saturation magnetisation is desirable in order to provide a strong magnetic flux and to facilitate the reading process. The useful coercive field represents a compromise; a high H_c prevents accidental erasure of data by surrounding magnetic fields and makes the reading step easier, but the writing processes more difficult. With the current recording heads, the desired value for the coercive field is in the 30–100 kA/m range for parallel recording.

The optimal material for magnetic recording media was, for many years, γ-Fe_2O_3 (maghemite). This form of Fe_2O_3, which has a defect-spinel crystal structure, can be prepared as small, elongated particles. The coercive field is produced within the required range by shape anisotropy (see Section 4.5.1). The process used to prepare small particles of

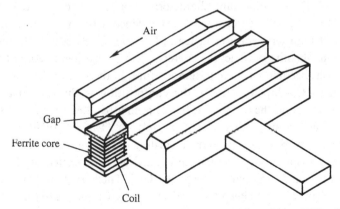

Fig. 5.18. Write/read head for hard-disc computer systems. (Adapted from White, 1985.)

maghemite takes advantage of the fact that hydrated ferric oxide, α-FeOOH, is easy to prepare as an elongated-particle powder. After dehydration to hematite, α-Fe$_2$O$_3$, the particles are reduced to magnetite, Fe$_3$O$_4$, in a H$_2$ atmosphere at 400 °C. This step results in a spinel structure. The last step is a reoxidation to γ-Fe$_2$O$_3$ at 250 °C. The particle dimensions vary from ~ 0.25 to 0.75 µm in length, and 0.05–0.15 µm in width, with a coercive field of 20–24 kA/m and a saturation magnetisation of 370 kA/m. Since the Curie temperature is one of the highest amongst the ferrites (~ 600 °C), the temperature dependence of all the magnetic properties is quite small.

Some competition was presented by tapes prepared from chromium oxide particles, CrO$_2$, which can be prepared with smaller dimensions and were expected to result in a higher storage density. The coercive field of CrO$_2$ is higher than that of maghemite: 40–80 kA/m (Mallinson, 1987). However, the Curie point of CrO$_2$ is a disadvantage since it is relatively low (128 °C), and leads to a higher temperature dependence of the magnetic properties.

Currently most of the tapes and discs are fabricated with a modified form of γ-Fe$_2$O$_3$. Due to a mechanism that is not yet fully understood, the addition of small quantities of Co to the γ-Fe$_2$O$_3$ particles results in a stronger coercive field, with values about 32 kA/m. This material, known as *Co-modified* γ-Fe$_2$O$_3$, or Co-adsorbed γ-Fe$_2$O$_3$, is obtained at relatively low temperatures ($T < 150$ °C). The increase in coercive field has been attributed to the formation of a uniaxial anisotropy resulting from the growth of a Co-rich surface layer, with a thickness about 1200 Å (Meng *et al.*, 1987), under the influence of the field created by the particle magnetisation (Kishimoto *et al.*, 1981). An important characteristic of this process is that all the magnetic properties show a weak dependence with temperature. If Co is added at high temperatures instead ($T > 300$ °C), it diffuses through the whole maghemite particle; the coercive field is also increased, but all the properties become sensitive to temperature. In the case of these materials, known as Co-doped γ-Fe$_2$O$_3$, the composition is thought to be close to CoFe$_2$O$_4$, and the increase in H_c has been attributed to the corresponding increase in magnetocrystalline anisotropy (Tachiki, 1960).

The fabrication of magnetic recording tapes and discs is relatively simple: a suspension of the magnetic particles is deposited on the plastic substrate, under an aligning magnetic field. The solvent is evaporated; a dry binder remains to keep the particles strongly attached to the substrate. The tape is finally rolled to obtain a compact coating.

Magnetic recording

Barium hexaferrite has magnetic and morphological characteristics which are extremely advantageous for a *perpendicular magnetic* recording medium. It can be obtained in the form of small, thin plates with the hexagonal axis normal to the plane, Fig. 5.19. Shape anisotropy favours an easy magnetisation direction within the plane; however, magneto-crystalline anisotropy is stronger and, for the usual particle dimensions, dominates the shape component. The magnetisation is therefore perpendicular to the plates. By preparing tapes (Fujiwara *et al.*, 1982), hard discs (Speliotis, 1987a) and flexible (or floppy) discs (Yamamori, Tanaka & Jitosho, 1991) with highly oriented particles, it has been possible to obtain storage densities significantly higher than with other recording media. Floppy discs with Ba ferrite as the magnetic medium have been prepared with a storage capacity of 4 MB (Yamamori, Tanaka & Jitosho, 1991). Some problems related to its commercial application remain (Speliotis, 1987b), but it is expected that the commercial application of perpendicular recording based on a Ba ferrite particulate medium will be developed in the near future.

5.4.5   *Magnetooptical recording*

In addition to the above 'inductive' technologies for information storage, devices based on the Faraday or Kerr effects are being developed. The aim is to increase two of the most appreciated characteristics in storage technology devices: data density and access time. The storage capacity of a magnetooptical disc is currently 600 MB (Jiles, 1991). An additional characteristic of these methods is that the wear problem is virtually eliminated.

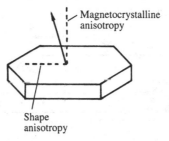

Fig. 5.19. A Ba hexaferrite platelet showing the easy directions for shape (horizontal) and magnetocrystalline (vertical) anisotropies.

The technique requires a thin ferromagnetic or ferrimagnetic film with uniaxial anisotropy, preferably with the easy direction perpendicular to the film plane. The film is initially saturated in the $+z$ easy direction. The information is written by means of an intense laser beam, focused in a small region of the film, Fig. 5.20(a). The light beam leads to rapid heating of this region above the Curie or compensation temperature; it is then left to cool again. If a small field (with a value below the coercive field of the film) is applied in the opposite direction, $-z$, during the whole process, the heated region adopts an opposite magnetisation direction, since when cooling down from the Curie (or compensation) temperature, the coercive field for that region is very small. In fact, the magnetisation inversion can be achieved without reaching the Curie point. The information 'bits' appear therefore as small domains with opposite magnetisation.

The reading of the information is performed by means of a polarised, low-intensity laser beam, by detecting the rotation angle of the transmitted (Faraday effect) or reflected (Kerr effect) beam, Fig. 5.20(b). In the case of reading by reflection, the magnetic film can be deposited on a reflecting

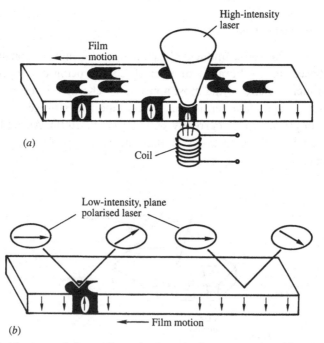

Fig. 5.20. Magnetooptical recording technology: (a) thermomagnetic writing; (b) reading by means of the Kerr effect. (Adapted from Marchant, 1990.)

substrate to facilitate the operation. The film is erased by saturation in a strong field in the $+z$ direction.

The optimal properties of the film include not only the magnetooptical properties, but also the thermal characteristics. The film should be thick enough to provide a strong, easily detected polarisation rotation, θ_f, but thin enough to prevent a strong light absorption. The writing energy is proportional to $C_v \Delta T / \alpha$ (Eschenfelder, 1970), where C_v is the heat capacitance, ΔT is the difference between the operation and Curie temperatures, and α is the optical absorption coefficient. A high T_c leads to a well-resolved 'bit', but also to a high writing energy. The optimal thickness is related with $1/\alpha$; since typical values of the absorption coefficient are in the 10^5 cm^{-1} range, film thicknesses are in the 500–1000 Å range. The writing beam is typically 5 µm in diameter.

GdIG, $Gd_3Fe_5O_{12}$, has a compensation point near room temperature which can be moved by Y substitution; it is a good candidate for magnetooptical memories. However, a problem with garnets in general is that their absorption coefficient is small, and they therefore require a higher energy for writing. Another candidate is bismuth iron garnet, $Bi_3Fe_5O_{12}$ (or 'BiIG'), because of its extremely high Faraday rotation coefficient (Gomi, Satoh & Abe, 1989).

5.5   *Microwave components*

The applications of ferrites at microwave frequencies (1–300 GHz) are based on electromagnetic wave propagation phenomena. Electromagnetic waves are formed by mutually perpendicular electric and magnetic fields in planes normal to the propagation direction, Fig. 5.21. Both the permittivity, ε, and the permeability, μ, are involved; the propagation velocity is $v = (\varepsilon\mu)^{-1/2}$ (Eq. (5.6)). The magnetic field vector in the plane XY, indicated as h_Y in Fig. 5.22, can be represented as formed by two components of equal magnitude, H^+ and H^-. As a function of time, these components rotate in opposite directions with angular frequency ω. Any propagating field in the Z direction can be analysed therefore in terms of these two oppositely rotating components of constant magnitude and angular frequency ω.

The application of a strong, saturating dc field parallel to the propagating direction $+Z$ reorients the spins of the sample and makes them precess about $+Z$, with a Larmor frequency ω_L. A microwave field of frequency ω_L is now applied to the sample along the $+Z$ direction; its effects can be analysed in terms of the two oppositely rotating fields H^+ and H^-. The precession movement is clockwise as observed in the direction $+Z$. The field H^+ rotates in the same direction as the spins; since the field frequency, ω_L, is the same as the precession frequency, resonance conditions are achieved. The component H^- rotates in an opposite direction with respect to the precession. Since resonance conditions are not achieved, the effective permeabilities of the ferrite to components H^+ and H^- are different. The components of the microwave field propagate with different velocities and wavelengths in the ferrite; as a result, at frequencies above the resonant frequency, the polarisation plane of the field rotates clockwise, and counterclockwise at frequencies below ω_L.

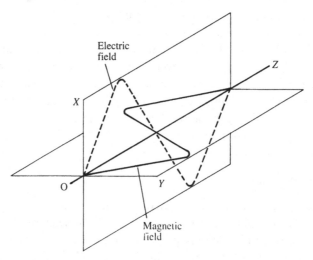

Fig. 5.21. Schematic representation of the electric and magnetic field amplitudes of a propagating electromagnetic wave. (Adapted from Brailsford, 1966.)

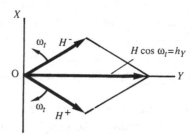

Fig. 5.22. Components H^+ and H^- of the magnetic field vector h_Y. (Adapted from Brailsford, 1966.)

Microwave components

Ferrite *isolators* (also known as Faraday rotation isolators) are typically employed to isolate source from load in microwave systems. The microwave power emitted by a 'klystron', or a 'magnetron', can be reflected from the load and cause severe damage to the source. The device used to prevent this reflection is made essentially of a ferrite rod, with a dc field parallel to the rod axis. The ferrite rod is placed between two absorbing elements, Fig. 5.23. These elements strongly attenuate the electric component of the wave, when its polarisation plane coincides with that of the element. The applied field and the ferrite properties are chosen in such a way that the polarisation plane of the microwave signal is rotated 45° when it propagates from the source to the load. Since the plane does not coincide with the orientation of the absorbing elements, the signal passes through without attenuation. A reflected signal travelling in the opposite direction is rotated a further 45° by the ferrite device. Since the total rotation is now 90°, the signal is strongly attenuated by the absorbing element.

Circulators are devices with three ports, where microwave signals can propagate only with a circulating pattern determined by the dc field, Fig. 5.24. A typical configuration includes a ferrite rod in the centre of the device, with the field perpendicular to it. The microwave propagation in the device is such that, if a signal is fed in port 1, the clockwise and the counter-clockwise components of the field reaching port 2 differ in phase by an integral multiple of 2π. The components arriving at port 3 have a phase difference of an odd multiple of π. When a signal enters

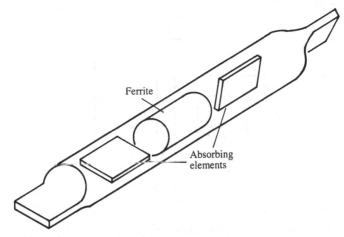

Fig. 5.23. Microwave isolator. (Adapted from Moulson & Herbert, 1990.)

215

port 1, it is transmitted to port 2 with little attenuation and is strongly attenuated with respect to port 3. Similar phenomena occur for signals entering port 2; they are weakly attenuated when travelling to port 3, but strongly damped towards port 1. The circulation sense is therefore $1 \rightarrow 2 \rightarrow 3 \rightarrow 1$. A circulator device allows the use of a single antenna for transmitting and receiving the reflected signal.

If a field with a value different from that for resonance conditions is applied perpendicularly to the axis of the ferrite, Fig. 5.25, the microwave signal is propagated with a phase change. *Phase shifters* are typically used in microwave antenna systems. These systems are formed of radiating dipoles; control of the phase shift between them allows accurate control of the emitted beam direction.

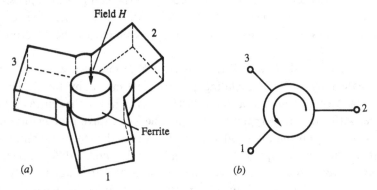

Fig. 5.24. (*a*) Microwave circulator and (*b*) its schematic representation. (Adapted from Moulson & Herbert, 1990.)

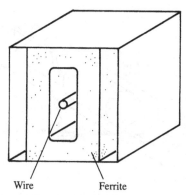

Fig. 5.25. Microwave phase shifter; the phase shift depends on the current flowing through the axial conductor. (Adapted from Soohoo, 1985.)

Other applications

Garnets have been extensively used in microwave devices because of their high resistivity. Since all of the Fe is maintained in the Fe^{3+} state, no electron hopping occurs. The saturation magnetisation can be widely varied by cation substitution in the basic garnet, YIG, between 8–140 kA/m. The Curie temperature is practically the same for all the rare-earth garnets; however, the temperature dependence around room temperature can be attenuated by Gd^{3+} substitution (Moulson & Herbert, 1990). Lithium ferrite, $Li_{0.5}Fe_{2.5}^{3+}O_4$, is the basis for many other microwave ferrites, because of its large saturation magnetisation (390 kA/m), high Curie temperature ($\sim 670\,^{\circ}$C) and high electrical permittivity. From these values, Li ferrite properties can be widely modified by cation substitutions, such as Co^{2+} and Ti^{4+} (Kuanr et al., 1988). The use of Mn–Zn and Ni–Zn ferrites for microwave applications is limited because of the losses associated with their relatively high electrical conductivity.

5.6   Other applications

Magnetic ceramics are used in a number of applications such as radar-signal absorbers, magnetic printers, magnetic levitation, lithiated materials for ionic conductivity, etc. A brief description of these applications is given below.

Electromagnetic wave *absorbers* in the GHz range are useful in microwave telecommunication technology to suppress electromagnetic interferences. A highly reflecting metallic structure disturbing radar navigation near an airport or a harbour can be strongly attenuated in the radar spectrum by such an absorber. Since the absorption is related to the Larmor resonance frequency in the ferrite, it is usually centred on a narrow frequency band. In these applications, a ferrite powder is typically dispersed in a resin and the mixture is applied to the structure as a paint. To obtain a wide-band absorber, a two-layer absorber with Co–Ti-substituted Ba hexaferrite, with a different composition in each layer, has been prepared (Aiyar et al., 1989). To increase the effective permittivity of the system, short metal fibres were interdispersed in the ferrite/resin mixture (Hatakeyama & Inui, 1984). The metal fibres worked essentially as dipole antennas, increasing the efficiency of the absorption.

In magnetic *printers*, the basic idea is to form a latent image by means of a magnetic drum and a system of magnetic heads. Each magnetic head produces a row of tiny magnetic dots on the drum; combined with the drum rotation, this step leads to the image to be printed. In the next step, the toner magnetic particles are attracted and fixed to the drum, to 'develop' the image. The transfer of the image to the paper is made by a pressure roller. This system would allow a printing speed of 6000 lines per minute, with a resolution of 120 dots per inch (Eltgen & Magnenet, 1980).

The use of ferrites and other permanent magnet materials for magnetic *levitation* has been investigated (Atherton, 1980). The permanent magnet was sandwiched between two thin Fe pole pieces and two coils to concentrate the magnetic flux. A calculation of the lift/weight and power/lift ratios showed that ferrite permanent magnets could be used in such an application.

Magnetic fluids can be used as a high-density solution for the sink-and-float separation of solids in suspension. This separation technique for non-ferrous metals is based on the anomalous viscosity increase of a magnetic suspension as a function of applied field. The magnetic fluid consists typically of magnetite particles with an average diameter of ~ 100 Å in water, with 15–25 wt% of ferrite. To prevent particle aggregation, a surfactant such as kerosene and oleic acid, or a polymer is usually

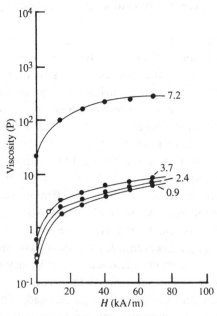

Fig. 5.26. Dependence of the viscosity of a magnetic fluid on the applied field; the volume concentration of magnetic particles is indicated (in vol%) (Tsutaoka *et al.*, 1989).

added. In experiments performed on a magnetic fluid prepared from ferrite particles recovered from mine drainage (Tsutaoka, Ema & Sato, 1989), a ten-fold increase in viscosity was observed at 64 kA/m, as compared with the zero field value, Fig. 5.26. A mixture of polyvinyl chloride, Al alloy and Cu scrap was separated by applying increasingly high fields; at 96 kA/m, 100% of the polyvinyl chloride was separated; at 120 kA/m, 83% of the Al alloy was recovered, and at 136 kA/m, all the Cu scrap and the remaining Al alloy were separated.

Li can be *inserted* into transition-metal oxides at room temperature in an inert atmosphere by a reaction between the oxide and n-butyl-Li. A low but non-negligible mobility of Li^+ ions has been observed in lithiated spinels (Goodenough *et al.*, 1984), which indicates that lithiated ferrites can be used as electrodes in ionic batteries. Lithiation in α-Fe_2O_3 (Thackeray, David & Goodenough, 1984), Fe_3O_4 (Fontcuberta *et al.*, 1986) and γ-Fe_2O_3 (Pernet *et al.*, 1989) has already been reported.

References

Abe, M., Itoh, T., Tamaura, Y. & Gomi, M. (1988). Ferrite–organic multilayer film for microwave monolithic integrated circuits prepared by ferrite plating based on the spray-spin-coating method. *Journal of Applied Physics*, **63**, 3774–6.

Aiyar, R., Rao, N. S. H., Uma, S., Rane, S. A. & Srivastava, C. M. (1989). Ba–Co–Ti based ferrite impregnated polyurethane paints as microwave absorbers. In *Advances in Ferrites: Proceedings of the Fifth International Conference on Ferrites, India, 1989*, Vol. 2. Eds C. M. Srivastava and M. J. Patni. Oxford & IBH Publishing Co. PVT Ltd, Bombay, pp. 955–60.

Arai, K. I., Yamaguchi, M., Ohzeki, H. & Matsumoto, M. (1991). Applications of YIG films to thin film inductors. *IEEE Transactions on Magnetics*, **27**, 5337–9.

Atherton, D. L. (1980). Maglev using permanent magnets. *IEEE Transactions on Magnetics*, **MAG-16**, 146–8.

Brailsford, F. (1966). *Physical Principles of Magnetism*. Van Nostrand, New York, p. 250.

Cantagrel, M. (1986). Development of ceramics for electronics. *American Ceramic Society Bulletin*, **65**, 1248–9.

Cullity, B. D. (1972). *Introduction to Magnetic Materials*. Addison-Wesley, Massachusetts, p. 589.

Eltgen, J. J. P. & Magnenet, J. G. (1980). Magnetic printer using perpendicular recording. *IEEE Transactions on Magnetics*, **MAG-16**, 961–6.

Eschenfelder, A. H. (1970). Promise of magneto-optic storage systems compared to conventional magnetic recording. *Journal of Applied Physics*, **41**, 1372–6.

Fontcuberta, J., Rodriguez, J., Permet, M., Longworth, G. & Goodenough, J. B. (1986). Structural and magnetic characterisation of the lithiated iron oxide $Li_xFe_3O_4$. *Journal of Applied Physics*, **59**, 1918–26.

Fujiwara, T., Issiki, M., Koike, Y. & Oguchi, T. (1982). Recording performance of Ba-ferrite coated perpendicular magnetic tapes. *IEEE Transactions on Magnetics*, **MAG-18**, 1200–2.

Globus, A. & Valenzuela, R. (1975). Influence of deviations from stoichiometry on the magnetic properties of Zn-rich Ni–Zn ferrites. *IEEE Transactions on Magnetics*, **MAG-11**, 1300–2.

Gomi, M., Satoh, K. & Abe, M. (1989). New garnet films with giant Faraday rotation. In *Advances in Ferrites: Proceedings of the Fifth International Conference on Ferrites, India, 1989*, Vol. 2. Eds C. M. Srivastava and M. J. Patni. Oxford & IBH Publishing Co. PVT Ltd, Bombay, pp. 919–24.

Goodenough, J. B., Thackeray, M. M., David, W. I. F. & Bruce, P. G. (1984). Lithium insertion/extraction reactions with manganese oxides. *Revue de Chimie Minerale*, **21**, 435–55.

Hanke, I. & Neusser, P. (1984). Power ferrites for high frequencies. *IEEE Transactions on Magnetics*, **MAG-20**, 1512–14.

Hatakeyama, K. R. & Inui, T. (1984). Electromagnetic wave absorber using ferrite absorbing material dispersed with short metal fibers. *IEEE Transactions on Magnetics*, **MAG-20**, 1261–3.

Jiles, D. (1991). *Introduction to Magnetism and Magnetic Materials*. Chapman and Hall, London, p. 331.

Kishimoto, M. K., Kitaoka, J., Andoh, H., Amemiya, M. & Hayama, F. (1981). On the coercivity of cobalt–ferrite epitaxial iron oxides. *IEEE Transactions on Magnetics*, **MAG-17**, 3029–31.

Kobayashi, K-I., Morinaga, H., Araki, T., Naka, Y. & Oomura, T. (1992). Low-loss Ni–Zn–Cu ferrite for deflection yoke. *Journal of Magnetism and Magnetic Materials*, **104–107**, 413–14.

Kuanr, B. K., Singh, P. K., Kishan, P., Kumar, N., Rao, S. L. N., Singh, Prabhat K. & Srivastava, G. P. (1988). Dielectric and magnetic properties of polycrystalline cobalt-substituted LiTi ferrites. *Journal of Applied Physics*, **63**, 3780–2.

Kugimiya, K. (1990). Ceramic materials for magnetic heads. *American Ceramic Society Bulletin*, **69**, 696–702.

Kugimiya, K., Hirota, E. & Bando, Y. (1974). Magnetic heads made of crystal oriented spinel ferrite. *IEEE Transactions on Magnetics*, **MAG-10**, 907–9.

Mallinson, J. C. (1987). *The Foundations of Magnetic Recording*. Academic Press, San Diego, p. 35.

Marchant, A. B. (1990). *Optical Recording; A Technical Overview*. Addison-Wesley, Massachusetts, p. 91.

Meng, R. L., Wang, Y. Q., Lin, C. S., Bensaoula, A., Chu, C. W., Hor, P. H. & Ignatiev, A. (1987). Study of H_c enhancement in Co-modified γ-Fe$_2$O$_3$ films. *Journal of Applied Physics*, **61**, 3883–5.

Moulson, A. J. & Herbert, J. M. (1990). *Electroceramics*. Chapman and Hall, London, p. 439.

Pernet, M., Rodriguez, J., Gondrand, M., Fontcuberta, J., Strobel, P. & Joubert, J. C. (1989). Lithium insertion into γ-Fe$_2$O$_3$. In *Advances in Ferrites:*

References

Proceedings of the Fifth International Conference on Ferrites, India, 1989,
Vol. 1. Eds C. M. Srivastava and M. J. Patni. Oxford & IBH Publishing Co.
PVT Ltd, Bombay, pp. 61–5.

Ruthner, M. J. (1989). Long term availability of spray roasted iron oxides for
the production of ferrites. In *Advances in Ferrites: Proceedings of the
Fifth International Conference on Ferrites, India, 1989,* Vol. 1. Eds C. M.
Srivastava and M. J. Patni. Oxford & IBH Publishing Co. PVT Ltd, Bombay,
pp. 23–34.

Sano, T., Morita, A. & Matsukawa, A. (1989). A new power ferrite for high
frequency switching power supplies. In *Advances in Ferrites: Proceedings
of the Fifth International Conference on Ferrites, India, 1989,* Vol. 1. Eds
C. M. Srivastava and M. J. Patni. Oxford & IBH Publishing Co. PVT Ltd,
Bombay, pp. 595–603.

Snelling, E. C. (1988). *Soft Ferrites, Properties and Applications.* Butterworths,
London, p. 294.

Snelling, E. C. (1989). Some aspects of ferrite cores for HF power transformers.
In *Advances in Ferrites: Proceedings of the Fifth International Conference on
Ferrites, India, 1989,* Vol. 1. Eds C. M. Srivastava and M. J. Patni. Oxford &
IBH Publishing Co. PVT Ltd, Bombay, pp. 579–86.

Soohoo, R. F. (1985). *Microwave Magnetics.* Harper and Row, New York, p. 204.

Speliotis, D. E. (1987a). Digital recording performance of Ba-ferrite media.
Journal of Applied Physics, **61**, 3878–80.

Speliotis, D. E. (1987b). Barium ferrite magnetic recording media. *IEEE
Transactions on Magnetics,* **MAG-23**, 25–8.

Stijntjes, Th. G. W. (1989). Power ferrites; performance and microstructure. In
*Advances in Ferrites: Proceedings of the Fifth International Conference on
Ferrites, India, 1989,* Vol. 1. Eds C. M. Srivastava and M. J. Patni. Oxford &
IBH Publishing Co. PVT Ltd, Bombay, pp. 587–94.

Tachiki, M. (1960). Origin of the magnetic anisotropy energy of cobalt ferrite.
Progress of Theoretical Physics, **23**, 1055–72.

Thackeray, M. M., David, W. I. F. & Goodenough, J. B. (1984).
High-temperature lithiation of α-Fe_2O_3: a mechanistic study. *Journal of Solid
State Chemistry,* **55**, 280–6.

Torii, M., Kobayashi, H. & Okuda, M. (1982). Study on application of radially
oriented ferrite magnet to stepper motor. In *Ferrites: Proceedings of the
Third International Conference on Ferrites, Kyoto, Japan, 1980.* Eds
H. Watanabe, S. Iida and M. Sugimoto, Center for Academic Publications,
Tokyo, pp. 370–4.

Tsutaoka, T., Ema, S. & Sato, H. (1989). Physical and chemical properties of
magnetic fluid using ferrite material from mine drainage and possibility for
some applications. In *Advances in Ferrites: Proceedings of the Fifth
International Conference on Ferrites, India, 1989,* Vol. 2. Eds C. M. Srivastava
and M. J. Patni. Oxford & IBH Publishing Co. PVT Ltd, Bombay,
pp. 1113–18.

Verweel, J. (1971). Ferrites at radio frequencies. In *Magnetic Properties of
Materials.* Ed. J. Smit. McGraw-Hill, New York, p. 98.

Application of ferrites

White, R. M. (1985). *Introduction to Magnetic Recording.* IEEE Press, New York.
Yamamori, K., Tanaka, T. & Jitosho, T. (1991). Recording characteristics for highly oriented Ba-ferrite flexible disks. *IEEE Transactions on Magnetics, 27,* 4960–2.

6   Other magnetic materials

In order to complete this review, a brief overview of magnetic materials other than oxides is presented in this chapter. Soft and hard metallic alloys are discussed first; instead of a detailed account of the numerous alloy systems, this overview focuses on the mechanisms for obtaining specific microstructures, which, in turn, lead to a precise control of *anisotropy* in soft materials, and to *coercivity* in hard materials. These discussions include examples of 'classic' alloys, as well as the recently developed soft amorphous alloys and the impressive supermagnets with extremely high coercive fields.

Due to their impact on information storage technology, a brief account of materials for magnetic recording media and heads is presented. Materials for magnetooptical recording are also discussed.

The next section deals with magnetic materials and applications that are only now emerging, but which arouse a strong and exciting interest, i.e. molecular magnets, magnetic imaging techniques and non-destructive evaluation by magnetic methods.

The last section is a short review of superconductors; their amazing magnetic properties, as well as two of their present applications are discussed finally. In each section, an effort has been made to indicate new directions in magnetic materials research.

6.1   Soft magnetic materials

The basic magnetic properties expected in soft magnets are *high initial and maximum permeabilities*, μ_i and μ_{max}, and *low coercivity*, H_c (lower than ~ 10 A/m). In many applications, soft magnets are subjected to ac fields; their frequency response and, particularly, their ac *losses* are therefore important.

The ratio M_s/K, where M_s is the saturation magnetisation and K is the total anisotropy (including magnetocrystalline, induced and shape aniso-tropy contributions), appears in most of the permeability relationships proposed to account for domain wall bowing, domain wall displacement and spin rotation mechanisms (as M_s^2/K, or $M_s^2/K^{1/2}$, depending on the model; see Section 4.4); therefore, this overview of soft ferromagnetic materials focuses on several mechanisms which may be used to maximise this ratio.

A number of the original references are old and sometimes not readily accessible; instead, some excellent reviews on soft magnetic materials are given (Chen, 1977; Chin & Wernick, 1980; Cullity, 1972). When expressing alloy compositions, an atomic per cent basis is used unless otherwise specified.

6.1.1   Crystalline alloys

Most soft ferromagnetic materials, other than ferrites, are metals and metallic alloys. When compared with ferrimagnetic oxides, metallic ferromagnets have several advantages: since atoms bearing magnetic moments are not 'diluted' by O atoms, saturation magnetisation values are considerably higher than in ferrites; atoms are closer and can establish a strong, *direct exchange* interaction with each other, resulting in a high Curie temperature. The value of the anisotropy constant can be drastically varied through several refined techniques, as discussed later in this chapter; consequently, metallic materials exhibit very high permeability values, as shown in Fig. 6.1, where the improvements in this property from 1900 to 1990 are plotted. In ferrites, a permeability value of 80 000 is exceptional; as a comparison, permeabilities up to 2×10^6 have been obtained in several metallic alloys.

As discussed in several sections of this book, electrical resistivity in ferrites can be 4–10 orders of magnitude larger than in metals; this difference explains the dominance of ferrites in most high-frequency applications. In metals, considerable efforts have been made to increase resistivity; however, due to the nature of metallic bonding, improvements by a factor of 2–3 are considered as exceptional.

One of the basic problems in 3d metals and alloys is how to account for the existence of *localised* magnetic moments in *free-electron* bands. Electrons in 3d metals do exhibit a localised character, as shown by the accurate fitting of the susceptibility of most ferromagnetic metals (in the paramagnetic state) to the Curie–Weiss law; on the other hand, non-integral values of the magnetic moment per atom at temperatures close to 0 K (see Table 4.5) are compatible only with an itinerant nature of the magnetic carriers.

The *band* theory of ferromagnetism (Slater, 1936; Morrish, 1965) provides a model to explain these conflicting properties. This theory is based on the assumption that spin-up and spin-down parts of the 3d band are unequally filled and couple by means of the exchange interaction. Electrons 'fill up' these bands by occupying the lowest energy levels first; in Fig. 6.2, these lowest energy states belong to the spin-up half-band. As 'filling' proceeds, electrons also begin to occupy the spin-down half band, until the Fermi level for that particular material is reached. The Fermi energy, E_f, is, of course, the same for both bands, since E_f is a fundamental property of the material. The 4s band is found in the same energy range of the density of states plot as the 3d band; however, both 4s half-bands are equally occupied and no net contribution to the magnetic moment appears.

This model qualitatively accounts for most of the apparently conflicting properties of metallic ferromagnets. The difference in occupancy between spin-up and spin-down half-bands leads to a spontaneous, localised magnetic moment per atom; the 'filling' process usually results in a

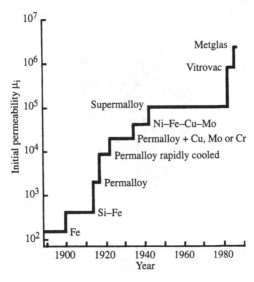

Fig. 6.1. Developments in high-permeability alloys from 1900 to 1990. (Adapted from Luborsky, 1980.)

non-integral magnetic moment per atom; electrons preserve their itinerant nature. Unfortunately, quantitative calculations on the band theory of ferromagnets are extremely complex.

When two (or more) 3d metals form an alloy (a solid solution), all alloying elements contribute electrons to a *common* band. If a 'rigid' band is assumed (in which the band shape is conserved over the whole composition range), the variation in the average number of electrons results in a linear variation of the magnetic moment between the end-members of the alloy. For most binary alloys, the magnetic moment as a function of the total number of electrons ($n = 3d + 4s$) falls on a straight line, increasing from Cr ($n = 6$) up to a maximum for $n \approx 8.3$ (between Fe and Co), then decreasing from this maximum down to Ca ($n = 11$), Fig. 6.3. Deviations from this behaviour have been explained in terms of a subdivision of each half-band and its effects on electron filling (Chikazumi, 1964).

Instead of giving a detailed description of all soft magnetic materials, this overview describes the basic mechanisms that can be used to modify

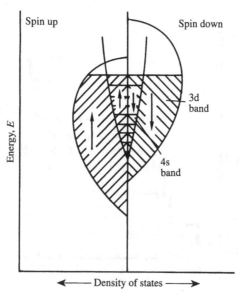

Fig. 6.2. Schematic representation of 3d and 4s energy bands with spin-up and spin-down half-bands in a 3d metal; 3d bands are separated by an exchange interaction.

Table 6.1. *Basic processes and
parameters used to modify
magnetic properties.*

Alloying
Order–disorder transformation
Directional order
Preferential orientation
Grain boundary effects
Surfaces
Shape

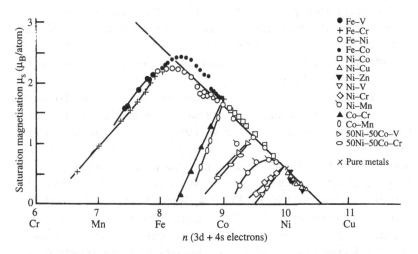

Fig. 6.3. Slater–Pauling curve: magnetic moment per atom as a function of the number of 3d + 4s electrons. (Adapted from Chikazumi, 1964.)

the properties of a given material. These mechanisms are listed in Table 6.1.

The properties of pure elements are rarely well adapted to applications. Also, in most cases, expensive processes are needed to purify the corresponding minerals in order to obtain the pure element. *Alloying* permits the 'tailoring' of materials to satisfy a specific application. Alloys allow variations of a given property often as a linear function of composition.

Fe is the basis of all soft ferromagnetic alloys, and is a good example of the ideas in the last paragraph. *Pure* Fe (>99.99 wt%) exhibits high saturation magnetisation (1.71×10^6 A/m) and Curie temperature (1043 K); its maximum permeability can attain $\sim 1 \times 10^6$, but impurities such as C, O, S and N, in concentrations as small as 200 ppm, can decrease μ to $\sim 10\,000$. A long and careful method (100 h at 1073 K in a H_2 atmosphere, followed by 3 h in vacuum at 1023 K) is needed to eliminate the impurities (Bozorth, 1951). Since these impurities enter the Fe lattice on interstitial sites, their effects can be attributed to the stress field (also known as the 'Cotrell atmosphere') they create. Domain wall displacement is hindered by this field and permeability is therefore drastically decreased. The effects on other properties are also important; yield strength, for example, doubles when 0.4% C is added to pure Fe (Chen, 1977a).

However, even if purifying techniques for Fe were inexpensive, other properties of the pure element, such as its electrical resistivity and corrosion resistance are inferior to those of the alloys. Of far greater importance is the fact that Fe alloys have an extremely wide range of properties and applications.

Fe–Si alloys (also known as electrical steels or silicon steels) are the most widely used soft magnetic materials. Si enters the Fe lattice in an interstitial position and leads to an increase in brittleness; saturation magnetisation decreases slightly as a simple consequence of dilution (Si is not magnetic). However, addition of Si has many advantageous effects. Pure Fe undergoes two transformations: α (bcc) $\leftrightarrows \gamma$ (fcc) 1185 K, and γ (fcc) $\leftrightarrows \alpha$ (bcc) at 1667 K, Fig. 6.4. For $2.5 <$ wt% Si < 6, the bcc structure is stable up to the melting point. This allows heat treatments (which may give rise to preferential orientation, as described below) at elevated temperatures without the limitation of the transformation. Anisotropy decreases to approximately half the value of pure Fe for ~ 5 wt% Si, resulting in an increase of μ. Finally, the electrical resistivity of Fe increases by a greater amount on addition of Si than of other alloying elements, Fig. 6.5; this, therefore, leads to reduced losses for ac applications.

Typical applications for electrical steel are in large distribution and power transformers, large generators and high-efficiency rotating machines. To reduce eddy-current loss, electrical steels are laminated into sheets which are usually 0.27–0.64 mm thick.

Fe–Al alloys are similar to Fe–Si alloys in some respects (high resistivity, hardness, permeability, loss). As with Si, addition of Al prevents the $\alpha \leftrightarrows \gamma$ transformation from occurring, and Fe–Al alloys can be grain-oriented by an appropriate process. One difference is that Al provides better corrosion resistance and ductility than Si; another difference is that the solubility limit of Al reaches ~30% Al (compared to ~2.5% in Fe–Si). The composition 75%Fe–25%Al, corresponding to the formula Fe_3Al, exhibits atomic order–disorder phenomena with changing temperature and these have interesting effects on the magnetic properties

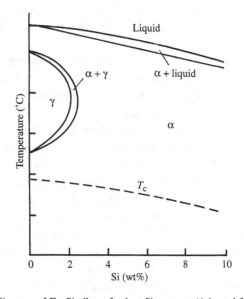

Fig. 6.4. Phase diagram of Fe–Si alloys, for low Si content. (Adapted from Stanley, 1963.)

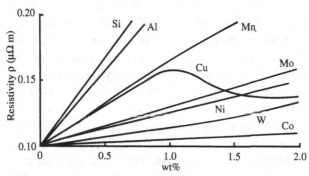

Fig. 6.5. Effect of alloying element on the electrical resistivity of Fe-based alloys. (Adapted from Bozorth, 1951.)

229

(discussed below). Alloys containing around 10% Al are sensitive to directional ordering; thus, permeability can be increased ~ 20 times by appropriate thermal treatment in a magnetic field. Fe–Al alloys, commercially known as *Alfenols* are suitable for high-temperature applications. A ternary alloy of composition 5.4%Al–9.6%Si–$\sim$85%Fe, known as *Sendust*, is mechanically hard and magnetically soft. Since it is too brittle to be laminated, it is used in either cast form or as sintered powder as a high-frequency inductor. Sendust thin films prepared by sputtering (Shibaya & Fukuda, 1977) are suitable for magnetic recording heads (see Section 6.3.2).

Practically all the remaining soft metallic ferromagnets are based on substitutional alloys of Ni–Fe and Co–Fe. Since at room temperature Fe is bcc, Ni is fcc and Co is hcp, the binary phase diagrams show a two-phase region in the middle of each diagram.

In the *Ni–Fe* system at room temperature, the α phase extends from 0 to $\sim$7% Ni, then $\alpha + \gamma$ mixtures from 7 to $\sim$50% Ni, and the γ phase from 50 to 100% Ni. γ-Phase alloys in the Ni–Fe system, known as *Permalloys*, exhibit a wide variety of magnetic properties, which may be controlled precisely by means of well-established technologies. Initial permeabilities up to 10^5 in an extremely wide temperature range, as well as coercive fields between 0.16 and 800 A/m, can be obtained (Chin & Wernick, 1980). Induced anisotropy of 65–85% Ni alloys can be drastically varied by field annealing and mechanical deformation (slip-induced anisotropy); an order–disorder transformation occurs for Ni_3Fe; finally, preferential orientation can be induced in 50%Ni–50%Fe.

Depending on the Ni content, there are three main groups of alloys: those with a composition of around 30% Ni exhibit the highest electrical resistivity of this system (*Rhometal, Invar*); those around 50% Ni exhibit high saturation magnetisation (*Hipernik*); in those around 80% Ni, the permeability reaches values of 10^5 (*Supermalloy* and *Mumetal*). Applications of Ni–Fe alloys range from power transformers and inductance coils to magnetic shielding and high-frequency transformers.

Fe–Co alloys with a composition near 35%Co–65%Fe (*Hiperco*) have the highest saturation magnetisation of all soft ferromagnets (1.95×10^6 A/m). Compositions close to 50%Co–50%Fe are also interesting since the maximum in permeability for this system is found in this range; also, additions of V allow a systematic variation in the coercive field from $\sim$2.4 kA/m for 49%Co–49%Fe–2%V (*Remendur*), to $\sim$16 kA/m for $\sim$44%Co–44%Fe–12%V (*Vicalloy*). The Hiperco alloys are used in

Table 6.2. *Order–disorder transformation types in soft ferromagnetic alloys.*

Alloy	T (K)	Crystal structure	
		Disordered (high temp.)	Ordered (low temp.)
FeCo	1003	bcc	$L2_0$
FePt	1570	fcc	$L1_0$
Ni_3Fe	733	fcc	$L1_2$
Fe_3Al	808	bcc	DO_3

Adapted from Chen (1977).

applications where the highest flux density is needed in the smallest volume (miniaturisation), such as in some aircraft generators. Remendur finds applications in devices where a high permeability is needed over a wide induction range, such as switching and storage coils.

Order–disorder transformations occur in a number of soft ferromagnetic alloys; there are four basic types of ordered crystal structures, exemplified by FeCo, FePt, Ni_3Fe and Fe_3Al, as listed in Table 6.2. The crystal structures of the ordered alloys are shown in Fig. 6.6. The bcc $\leftrightarrows L2_0$ transformation, Fig. 6.6(*a*), involves a simple crystallographic ordering with alternating positions. In the ordered structure, Fe atoms occupy the corner positions and Co atoms the body-centred sites (or vice versa) of the cubic unit cell. The other type of transformation involving a bcc lattice, the bcc $\leftrightarrows DO_3$ transformation observed in Fe_3Al, needs eight original bcc cells to be described adequately, Fig. 6.6(*d*). The transformations involving fcc disordered lattices, FePt and Ni_3Fe are shown in Figs. 6.6(*b*) and (*c*), respectively. Note that the $L1_0$ ordered structure is represented here by a hypothetical 'face-centred tetragonal' ('fct') lattice. Although the correct description is body-centred tetragonal (bct), the fct representation allows a clear visualisation of the change from the original fcc structure.

As in all transformations, the kinetics (the rate at which the equilibrium phase is reached) is critical. Since the magnetic properties of the disordered phase are higher than in the ordered alloy, the formation of the latter is minimised by quenching from the high-temperature, disordered state. The important parameter is the 'critical quenching rate', i.e. the slowest cooling rate which allows the disordered structure to be retained (metastably) to room temperature. The ordered $L2_0$ structure is particularly stable and forms rapidly; the critical quenching rate in FeCo is too high, therefore, and the disordered phase is rarely obtained. For this reason, V is added, since it leads to a slowing down in the transformation kinetics.

The most important consequences of atomic disorder are observed in the permeability value, shown in Fig. 6.7 for Ni–Fe alloys. Permeability peaks at $\sim 75\%$Ni ($\sim Ni_3Fe$) for quenched (disordered) samples; for slowly cooled (ordered) samples, two small maxima are observed. These results can be explained in terms of the anisotropy behaviour, Fig. 6.8. Quenched samples show a monotonic decrease in anisotropy as the Ni content increases, crossing the $K_1 = 0$ axis for $\sim 75\%$ Ni, the composition with maximum permeability. Slowly-cooled samples exhibit similar behaviour overall, but anisotropy crosses the $K_1 = 0$ axis at 65% Ni and shows a minimum for Ni_3Fe. Accordingly, the permeability of

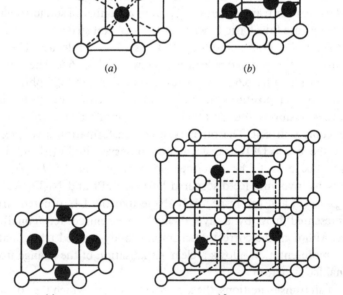

Fig. 6.6. The four basic ordered lattices observed in order–disorder transformations in ferromagnetic alloys: (*a*) the $L2_0$ lattice; (*b*) the Ll_0 lattice; (*c*) the $L1_2$ lattice; and (*d*) the DO_3 lattice. (Adapted from Chen, 1977.)

ordered alloys exhibits two small maxima, as shown in Fig. 6.7: one at $\sim 85\%$ Ni, corresponding to a minimum in $|K_1|$, and the other at $\sim 60\%$ Ni, where K_1 changes sign. The permeability variation for $K_1 \approx 0$ is not as drastic as the one for the disordered alloys because the magnetostriction constant, λ, is relatively far from zero (λ is not sensitive to order–disorder). In order to maximise permeability, the following procedure is adopted: first a composition with $\lambda \approx 0$ is chosen; the condition $K_1 \approx 0$ is then obtained by appropriate thermal treatment.

In disordered structures, the electrical resistivity is usually larger than in the corresponding ordered phase. This effect has a simple origin: an ordered structure presents a modulated scattering potential to conduction electrons and electrons having appropriate wave functions can easily diffuse through such an ordered potential system. Consequently, their resistivity is usually low. In a disordered structure, however, electrons face a disordered array of different scattering centres. The mean free path (the average distance between two consecutive scattering processes) is smaller

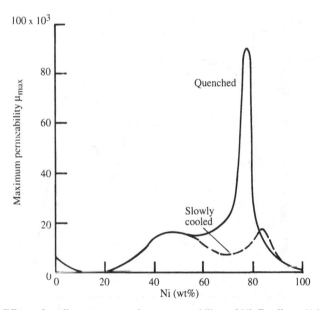

Fig. 6.7. Effects of cooling rate on maximum permeabilitys of Ni–Fe alloys. (Adapted from Bozorth, 1953.)

and resistivity increases. In Ni–Fe alloys, resistivity increases from ∼0.12 to ∼0.18 μΩ m when the alloys become disordered (Bozorth, 1951).

Mechanical properties of disordered alloys are also different to those of ordered alloys. This can have a bearing on the techniques used to prepare the alloys. Ordered structures are usually harder than disordered ones. In the former, dislocations have higher energy; the Burgers vector is larger because it is defined on the basis of the superlattice. Also, dislocation movement is hindered by antiphase domain boundaries which may be present in the ordered state.

Directional order (also known as 'magnetic annealing' or 'field annealing'), as briefly described in Section 4.4.2, can appear when the energy of a particular short-range arrangement (an atomic pair, for instance) depends on the relative orientation of that arrangement with respect to the magnetisation direction. An important condition for this to occur is that diffusion should be non-negligible at temperatures below the Curie temperature. The effects of directional order on permeability can be drastic; in the largest difference reported, maximum permeability in a

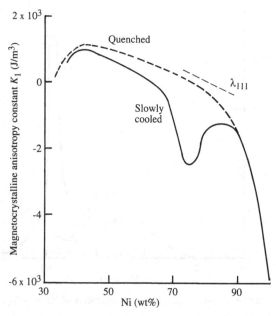

Fig. 6.8. The magnetocrystalline anisotropy constant as a function of the cooling rate in Ni–Fe alloys. (Adapted from Bozorth, 1953.)

6.5%Si–Fe single crystal increased from 5×10^4 to 3.8×10^6 after heat treatment in a magnetic field (Goertz, 1951). The effects of several thermal treatments on hysteresis loops of a 65%Ni–35%Fe alloy are shown in Fig. 6.9. These phenomena have been observed in many alloys, such as Fe–Al, Ni–Fe, Fe–Si, etc., as well as in ferrites (see Section 4.4.2).

Directional order can be visualised in a two-dimensional lattice of a hypothetical AB alloy as shown in Fig. 6.10. Both Figs. 6.10(*a*) and (*b*) show a disordered alloy (disordered in the sense that there are as many

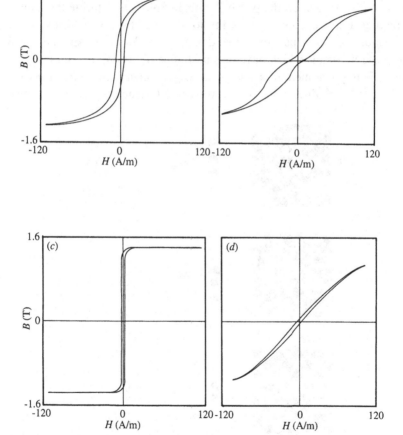

Fig. 6.9. Effect of heat treatment on hysteresis loops of a 65%Ni–35%Fe alloy, annealed at 1000 °C: (*a*) rapidly cooled; (*b*) slowly cooled; (*c*) cooled in a longitudinal field; and (*d*) cooled in a transverse field. (Adapted from Bozorth, 1951.)

235

Other magnetic materials

AB, AA and BB pairs as in a simple statistical distribution); the only difference between them is that in Fig. 6.10(*b*) there are more like-atom pairs (AA and BB) oriented along the vertical axis than in the horizontal direction; in Fig. 6.10(*a*) the number of like-atom pairs is nearly the same for any direction. In the case of a long-range ordered alloy, Fig. 6.10(*c*), there is no possibility of directional order; all atom pairs are of the type AB.

The directional order of Fig. 6.10(*b*), for example, can be produced by annealing at $T < T_C$ in a vertical field (assuming the energy of AA pairs is lowered when they are oriented parallel to the magnetisation). A field of 800 A/m is usually enough to produce these effects. The applied field has no direct effect on the pair ordering; its role is to saturate the sample, creating a single domain. It is the local magnetisation which leads to directional order. This implies that directional order occurs spontaneously in those materials which have non-negligible diffusion at room temperature. This phenomenon is known as *self-magnetic annealing*, or *ageing*. The effects of directional order on crystal structures have not been

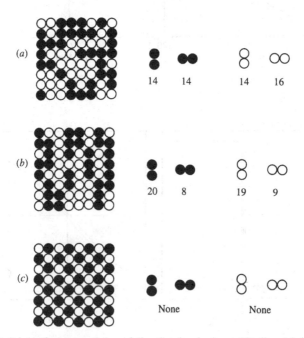

Fig. 6.10. Schematic representation of directional order in an AB alloy: (*a*) disordered alloy, where like-atom pairs are ordered in no particular directions; (*b*) directional order; and (*c*) long-range ordered alloy (no like-atom pairs). (Adapted from Graham, 1959.)

observed directly by x-ray or neutron diffraction. It has been estimated (Slonczewski, 1963) that the fraction of oriented like-atom pairs is $<1\%$, which is too small to be detected by the usual diffraction methods. However, its effects on magnetic properties can be considerable.

Directional order results in a strong induced anisotropy (and an easy axis) in the field direction; this leads to large 180° domains separated by highly mobile domain walls parallel to this axis. Induced anisotropy values in the range. $100-400\,\mathrm{J/m^3}$ have been observed in Ni–Fe alloys (Ferguson, 1958). The application of a field parallel to the induced anisotropy axis results in extremely soft behaviour (Fig. 6.9(*c*)); saturation is reached for small fields, as a result of domain wall mobility. If the field is instead applied in a transverse direction, completely different behaviour is observed, Fig. 6.9(*d*). Since domain wall displacement has no net effect on the magnetisation in the field direction (domain walls are 180° Bloch walls, oriented in a direction perpendicular to the field), magnetisation proceeds mainly by spin rotation (see Section 4.5.1). The hysteresis loop in Fig. 6.9(*a*) was obtained after heat treatment at $T > T_C$ (T_C for this composition is around 900 K), and therefore corresponds simply to the stress-free state.

When the material is annealed at $T < T_C$ with no field (or slowly cooled from $T > T_C$ with no field), Fig. 6.9(*b*), a 'constricted' loop is observed. These hysteresis loops, as well as the decrease in permeability usually observed in these cases, can be explained by considering the effect of directional order on domain walls. In domain walls, the magnetisation vector rotates by a small angle from one atom to the next; a full 180° rotation takes place over some 50 interatomic distances. If sufficient diffusion occurs to allow substantial atomic movements, it can be expected that directional order will be established at the domain walls. The main difference, as compared to directional order within domains, is that like-atom pairs form complex arrangements, tending to follow the magnetisation orientation within the wall. When a field is applied to such a material, not only is an anisotropy with a different easy axis induced in each domain, but domain walls become *pinned* on their original sites (which were established during heat treatment). An additional barrier has to be overcome to cause displacement of domain walls. This barrier remains physically located at the original domain wall positions; domain walls tend therefore to become pinned each time they pass through their original position. The hysteresis loops of Fig. 6.9(*b*) are characteristic of self-magnetic annealed materials; when annealing at $T < T_C$ leads to a decrease (a 'degradation') in magnetic properties, this is sometimes known as *ageing*.

Slip-induced anisotropy, or deformation-induced anisotropy is a phenomenon which also leads to directional order. It has been observed in a number of alloys, and especially in equiatomic Ni–Fe compositions, after cold-rolling. The formation of like-atom pairs occurs by slip, as shown in Fig. 6.11 for an AB alloy. Slip-induced anisotropy creates a hard axis in the rolling direction in Ni–Fe alloys; in other materials, the rolling direction becomes the easy axis. Another important feature is that directional order increases with the degree of long-range order prior to the rolling step. The magnitude of the induced uniaxial anisotropy can be as high as 50 times that obtained by means of thermal annealing (Chin & Wernick, 1980). Finally, when deformation is severe (a reduction in thickness of more than $\sim 50\%$), the structure becomes highly disordered (slip systems intersect each other) and induced anisotropy decreases.

Magnetic properties can also be strongly modified by a *preferential orientation* of grains (in many texts, it is known simply as *texture*).

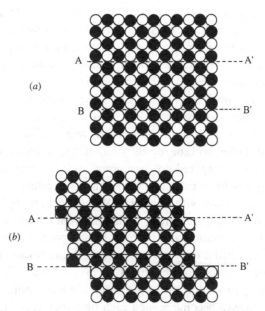

Fig. 6.11. Schematic representation of directional order induced by slip in an AB alloy; (a) before slip; (b) after slip. Note that a slip involving two interatomic distances (plane BB') has no effect on the directional order. (After Chin, 1971.)

Table 6.3. *Fabrication steps in oriented steels.*

(1) Melting of raw materials; casting into ingots.
(2) Hot rolling in two steps at ~ 1600 K, to 1.5–2.5 mm thickness.
(3) Cold rolling in two steps to 0.25–0.40 mm thickness with an intermediate annealing at 1000–1250 K.
(4) Annealing at ~ 1050 K in wet H_2 (primary crystallisation).
(5) Annealing at 1450 K in dry H_2 (secondary crystallisation).

Soft ferromagnetic materials for ac applications are usually prepared in the form of thin laminated sheets (essentially to decrease eddy-current losses; see the end of this section). A proper combination of rolling and annealing processes can lead to a preferential orientation of grains, which results in extremely favourable magnetic properties. The fabrication steps for electrical steels are listed in Table 6.3.

During the initial hot-rolling process, slabs of few centimetres thick are transformed into sheets ~ 2 mm thick. The final thickness (0.25–0.40 mm) is achieved through a two-step, cold lamination technique. This severe deformation results in the creation of defects (mainly dislocations) and internal stresses in the sheet; thermal treatments at the end and between the lamination stages are needed to restore the stress-free condition to grains and to obtain the required soft magnetic properties.

Annealing of cold-rolled material can lead to several phenomena, depending on the annealing temperature. For temperatures less than half the melting point ($T < T_m/2$), stress relief takes place by polygonisation, point defect migration, and plastic deformation. During polygonisation, edge dislocations of the same sign realign and result in small-angle subgrain boundaries; screw dislocations of opposite sign mutually annihilate. Point defects become mobile as a result of thermal activation; they annihilate, recombine and rearrange to eliminate stress. As the temperature rises, some plastic deformation can occur in grains where the residual stress overcomes the elastic limit (which, for most materials, decreases with temperature).

Annealing at $T > T_m/2$ leads to *recrystallisation*. At high temperatures, new grains can be nucleated and grown, not only relieving residual stress, but completely replacing old grains. In *primary* recrystallisation, grain growth occurs at a 'continuous' rate resulting in a uniform coarsening of the microstructure. If the temperature is further increased, *secondary* recrystallisation takes place; grain growth occurs by a 'discontinuous' process, where only a few grains grow at an abnormally high rate at the expense of all the others. Secondary recrystallisation can therefore be used to obtain unusually large grains with a preferential orientation. This process can be understood on the basis of the difference in surface energy between different crystal planes. Since grains can grow to a size comparable to the final sheet thickness, their surface energy becomes significant. Electrical steels, where strong preferential orientation has been obtained, have the bcc, α-Fe crystal structure. The highest-density planes are $\{110\}$ as shown in Fig. 6.12, and therefore they tend to remain within the sheet plane. Additionally, directions $\langle 100 \rangle$ are aligned parallel to the rolling direction. Preferential orientation is usually indicated as $(hkl)[uvw]$, where (hkl) is the rolling plane and $[uvw]$ the rolling direction. In this case of electrical steels, the preferential orientation is $(110)[001]$ (also known as 'cube-on-edge'), Fig. 6.13(a).

Magnetic properties of electrical steels are greatly improved by preferential orientation. Since the easy direction in Si–Fe is $\langle 100 \rangle$, oriented sheets are extremely soft in the rolling direction: permeability

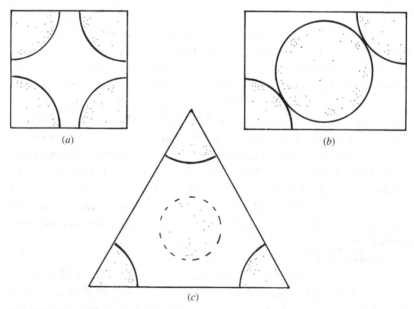

(a) (b)

(c)

Fig. 6.12. Principal crystal planes of the bcc structure of iron; (a) $\{100\}$; (b) $\{110\}$; and (c) $\{111\}$. Planes with highest atomic density are $\{110\}$.

increases and hysteresis loss and coercive field decrease. However, oriented materials are 'harder' when the field orientation is not parallel to the rolling direction, because any other direction contained in the sheet plane (such as $\langle 110 \rangle$ or $\langle 111 \rangle$) is harder than [100]. Oriented steels are therefore suitable for applications where the field direction is essentially unchanged, such as transformers. For applications in which the field direction varies with time (rotors or stators in motors and generators, for instance), an orientation (100)[001] is better, since the sheet plane contains two easy directions, Fig. 6.11(*b*). This preferential orientation has been achieved in Si–Fe, mainly through two techniques: a cross-rolling step (the second rolling is performed at 90° to the first rolling direction) and by modifying the surface energy of crystal planes (by means of an Al_2O_3 coating during the final anneal). However, these techniques are more expensive than the original orientation process and have not become a serious competitor to (110)[001] oriented sheets.

Energy *loss* is an extremely important subject in soft ferromagnetic materials, since the amount of energy wasted on processes other than magnetisation can prevent the ac applications of a given material. Fig. 6.14 illustrates the improvements in energy losses achieved between 1900 and 1980. The first decrease in losses, from 1900 to ~ 1930, resulted from advances in metallurgy: alloying, hot-rolling techniques, impurity control, etc. The second reduction (after ~ 1935) is associated with the development of preferential orientation technology.

Losses can be divided into two general groups: hysteresis and eddy-current loss. The former is related to the area within a hysteresis loop, i.e., to the energy needed to displace domain walls through the lattice.

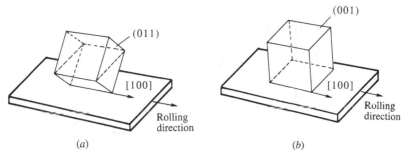

Fig. 6.13. Preferential orientations by rolling and annealing methods: (*a*) (110)[001], or 'cube-on-edge' preferential orientation; (*b*) (100)[001], or 'cube-on-plane' preferential orientation.

Other magnetic materials

The main energy loss is attributed to the creation and annihilation of the domain wall surface during its displacement (Guyot & Globus, 1976). Hysteresis loss decreases by decreasing the coercive field, by means of the various mechanisms discussed in this section. An approximate, general expression for hystersis loss as a function of frequency is simply:

$$P_h = W_h f \qquad (6.1)$$

where P_h is the hysteresis loss, W_h is the area of the loop under dc conditions and f is the frequency.

As briefly discussed in Section 4.7, a time-varying magnetic flux leads to a time-varying electric potential (due to Faraday's law of electromagnetic induction, Eq. (4.71)). Eddy-current loss is a function of Bf^2/ρ (Eq. (4.69)), where B and ρ are the magnetic flux density and the resistivity, respectively. Since B is directly related to the permeability, eddy-current loss is expected to be particularly high in ferromagnetic metallic alloys, where μ is large and ρ rather small. The total loss can be written:

$$P_T = W_h f + kf^2 \qquad (6.2)$$

where k is constant for a given material and geometry. Experimental results, however, are not in agreement with Eq. (6.2); an additional contribution must be present that increases as frequency increases. This 'anomalous' loss is due to the fact that Eq. (4.69) was obtained on the

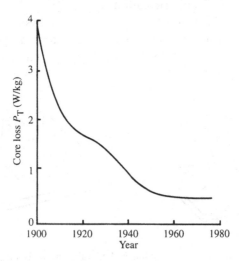

Fig. 6.14. Improvements in core loss at 1 T of induction level (60 Hz) between 1900 and 1980. (Adapted from Chin & Wernick, 1980.)

assumption of a uniform magnetised state. On a local scale, domain structures lead to an extremely inhomogeneous magnetisation which varies from $+M_s$ to $-M_s$ across a domain wall. For a realistic calculation, eddy currents should be evaluated taking into account the detailed geometry of the particular domain structure associated with the active magnetisation mechanism; i.e., whether it involves domain wall bowing, wall displacement and/or spin rotation ('micro-eddy' currents). This is obviously an extremely complex problem. To make things worse, there is experimental evidence that, for high-flux rates, the number of domain walls increases with frequency (Houze, 1967). To produce a higher flux change as the field rate increases, it is easier to create additional domain walls than to displace the existing ones at a higher velocity. Microeddy currents effectively limit domain wall velocity.

Eddy-current loss can be limited by increasing resistivity through alloying, as discussed in this section. Other methods include lamination and creation of insulating grain boundaries (limiting eddy-current paths), as well as producing a fine domain structure by induced anisotropy; eddy currents are lower for small displacements of many walls, than for large movements in coarse domain structures.

Some developments in grain-oriented Si–Fe alloys have used grain-growth inhibitors to enhance preferential orientation during secondary recrystallisation (Taguchi *et al.*, 1976). The features of domain structures in grain-oriented Si–Fe have been studied by Nozawa *et al.* (1988).

6.1.2 *Amorphous alloys*

Long-range magnetic order in atomically disordered materials leads to new and exciting phenomena and applications. It might seem contradictory to have domains extending a few cubic microns in media where atoms (the magnetic moment carriers) are spatially disordered. The terms 'amorphous' and 'disorder' are used rather loosely, however; there are many types of disorder. For present purposes, the model closest to an amorphous alloy (also known as 'a metallic glass') is a random packing of hard spheres, with a strong tendency to maintain a constant separation between nearest neighbours.

The density of amorphous materials is just a few per cent lower than that of crystals of the same composition; the short-range order in both is expected to be comparable. An illustrative example of the occurrence of short-range order in an amorphous material is provided by silicate glasses, where O tetrahedra are almost always present as the main structural building blocks. Crystalline and amorphous arrangements differ in that, in the latter, tetrahedra are disordered. To date, however, no evidence for short-range order has been reported in metallic glasses.

As with most amorphous phases, metallic glasses present the so-called *glass transition*, T_g. This temperature is usually considered as the point where (on heating) a glass becomes closer to a liquid; it is characterised by an abrupt change in most of the macroscopic properties. The drop in viscosity (by several orders of magnitude) is one criterion used to define T_g. For most metallic glasses, the crystallisation temperature is close to the glass temperature, which makes the study of this transition difficult. An exception is $Pd_{40}Ni_{40}B_{20}$ (Ström-Olsen *et al.*, 1991); the amorphous phase was stable enough to allow a study of T_g.

The existence of amorphous ferromagnetic alloys (first prepared by Duwez (1967)) shows that magnetic exchange is an essentially nearest-neighbour interaction. The spins of neighbouring atoms can become ordered if they are close enough, regardless of their spatial distribution, with Curie temperatures as high as 800 K. This is a consequence of the tendency to conserve the electronic structures in metals; energy bands of a molten metal have many common features with bands in the solid. As in crystals, ferro-, antiferro-, ferrimagnetic and other complex spin arrangements have been observed in amorphous materials.

The lack of periodicity makes any description of the *atomic arrangement* in amorphous solids extremely difficult. This contrasts with crystalline phases, where long-range order allows an accurate description of the spatial distribution of huge quantities of atoms, from a knowledge of just a few (the unit cell). X-ray or electron diffraction experiments on amorphous materials usually lead to an intense, ill-defined ring around the transmitted beam, with some other dim, blurred concentric rings, Fig. 6.15. The intense ring is associated with the nearest-neighbour interatomic distance; second- and higher-order neighbour distance vary considerably, resulting in the diffuseness of subsequent diffraction rings.

Amorphous solids can be characterised to some extent by means of the *radial distribution function*, $RDF(r)$ (Luborsky, 1980), which describes the average number of atoms between r and $(r + dr)$, with the origin taken at any particular atom. The average number of atoms is then $4\pi r^2 P(r)\, dr$, where $P(r)$ is the atomic distribution obtained from neutron or x-ray experiments:

$$RDF(r) = 4\pi r^2 P(r)\, dr \qquad (6.3)$$

A *pair correlation* function is sometimes used, which is related to the atomic distribution function by:

$$PCF(r) = P(r)/P_0 \qquad (6.4)$$

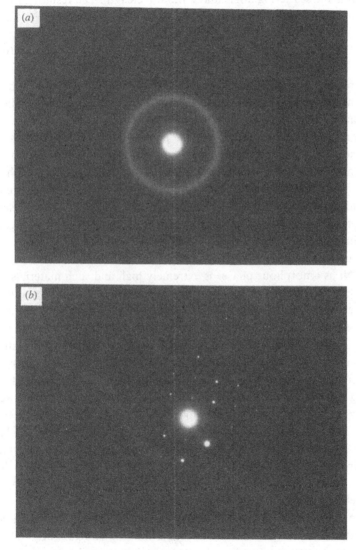

Fig. 6.15. Electron diffraction patterns of: (*a*) an amorphous Fe–Zr thin film, and (*b*) the same thin film after crystallisation at high temperature. (Valenzuela, unpublished results.)

245

where P_0 is the average atomic density. The radial distribution function can be obtained from the *interference* function, $I(K)$, where K is evaluated from scattered intensities by:

$$K = 4\pi \sin \theta/\lambda \qquad (6.5)$$

where θ is the scattering angle and λ the wavelength. The resulting radial distribution function is then:

$$RDF(r) = 4\pi r^2 P_0 + (2\pi/r) \int_0^\infty K[I(K) - 1] \sin(Kr)\,dK \qquad (6.6)$$

Interference and radial distribution functions for $Fe_{80}P_{13}C_7$ amorphous alloys are shown in Fig. 6.16 (Note: compositions in amorphous alloys are usually given as atom percentages).

Amorphous materials are *metastable phases*, since the crystalline state corresponds to the thermodynamic equilibrium state. To obtain an amorphous material, a drastic cooling from the liquid or the gas phase down to a temperature where diffusion is negligible has to be performed to avoid crystallisation, Fig. 6.17. Experimentally, a number of curves such as (a) and (b) are obtained as a function of the cooling rate, which suggests that different disordered atomic arrangements may exist.

The cooling rate required to prepare elemental 3d metals such as Fe, Co or Ni as amorphous phases is extremely high and such materials are unstable even at room temperature; *alloying* is therefore essential to obtain a stable solid. There are two general groups of alloys: transition metal–metalloid (TM–M), and rare earth–transition metal (R–TM). In the first group, a metalloid such as B, Si, C, P, acts as 'glass former', i.e., an atom that stabilises the amorphous phase by decreasing diffusion rates; an additional effect is to lower the melting temperature, which makes the cooling process easier. Eutectic compositions (which possess the lowest melting point) are particularly important for glass formation. In the second group, the rare-earth atoms presumably play the role of glass formers. Typically, the magnetic order in the first group is ferromagnetic, but ferrimagnetic in the second. Since the former usually leads to higher Curie temperatures, saturation magnetisations and permeabilities, properties useful in applications, this overview focuses on them.

Since amorphous alloys are prepared by quenching, the need for heat extraction at a high rate imposes a limiting size (about $\sim 60\ \mu m$ maximum) on at least one dimension. Amorphous samples are usually in the form of thin films, discs or ribbons. They can be prepared from the

246

vapour phase by methods such as vacuum deposition and sputtering. To obtain an amorphous phase, deposited atoms are prevented from diffusing by a low substrate temperature. Electrodeposition from electrolytic baths can also be used to obtain amorphous coatings.

Methods based on quenching from the *melt*, such as piston and anvil, double piston, torsion catapult and roller casting (Luborsky, 1980) permit larger quantities of amorphous materials to be obtained. The last of these, also known as *melt spinning*, is probably the method most used to obtain amorphous ribbons of thickness ~ 10–$60\,\mu m$, width up to $\sim 17\,cm$ and virtually infinite length. In this method, the molten alloy is propelled through a small hole or a slot onto a massive, cold metallic disc rotating

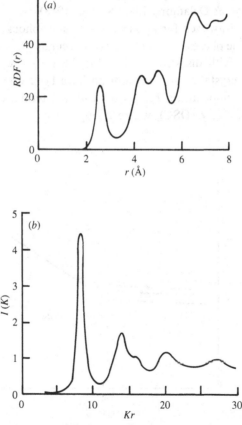

Fig. 6.16. (*a*) Radial distribution function, $RDF(r)$, and (*b*) interference function, $I(K)$, for a $Fe_{80}P_{13}C_7$ amorphous alloy. (Adapted from Cargill, 1975.).

at high speed, Fig. 6.18. The molten jet is rapidly solidified (rates of 10^6 K/s can be obtained) and emerges as a solid ribbon from the wheel. The quenching rate depends on the wheel speed, particularly on the speed of the wheel surface with which the molten metal comes into contact (the tangential speed); a typical figure for the critical tangential speed to obtain a completely amorphous phase is ~ 30 m/s. In spite of the small thicknesses, ribbons often exhibit inhomogeneities due to the slight difference in quenching rate between the wheel side (highest quenching rate) and the air side. In some cases, crystallites have been detected on the air side of an otherwise amorphous ribbon (Escobar *et al.*, 1991). Surface crystallisation, however, has also been detected on the wheel side of the ribbon and has been attributed to nucleation at the wheel–ribbon interface, at sites such as microscopic gas pockets, trapped during the preparation process (Gibson & Delamore, 1988; Köster, 1988).

A potential drawback for applications of amorphous ribbons is that, being metastable phases, *crystallisation* can occur as the working temperature increases, with disastrous effects for the magnetic properties. For these reasons, crystallisation phenomena have been widely studied. The crystallisation temperature, T_x, can be readily determined by differential scanning calorimetry (DSC), where it appears as a strong, exothermic

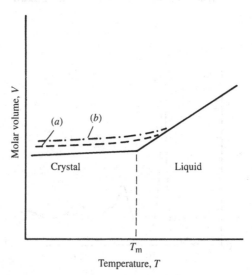

Fig. 6.17. Schematic variation of molar volume as a function of temperature for a liquid → crystal and a liquid → amorphous transition; cooling rate (*b*) is higher than (*a*).

peak. T_x is not constant for a given composition, but depends on the heating rate, since crystallisation from the amorphous phase is an essentially kinetic phenomenon. Crystallisation results are usually analysed by means of the Avrami-based model of nucleation and growth. The crystallised fraction, x, at time t, is given by:

$$x = 1 - e^{-Kt^n} \tag{6.7}$$

where K is an Arrhenius-type rate constant ($K = K_0 e^{-E/RT}$; E is the activation energy and R the gas constant) and n is the Avrami exponent; the latter can be partitioned (Ranganathan & von Heimendhal, 1981) according to:

$$n = a + bp \tag{6.8}$$

where a is associated with the nucleation rate ($a = 1$ for constant-rate nucleation and 0 for pre-existing nuclei), b depends on the dimensionality of the growing phase (1, 2 or 3), and p is related to growth mechanism ($p = 0.5$ for diffusion-controlled growth and 1 for interfacial growth). A non-isothermal expression relating E and n has been derived (Marseglia, 1980):

$$\frac{d \ln(r/T)}{d(1/T)} = -\frac{E}{nR} \tag{6.9}$$

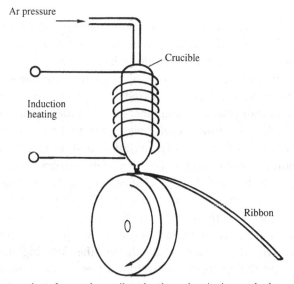

Fig. 6.18. Preparation of amorphous alloys by the melt-spinning method.

where r is the heating rate. This expression allows evaluation of both E and n from DSC experiments. Similar results can be obtained by using another transformation model, the Kissinger expression; however, in spite of the similarity in numerical results, this approach is based on a physically different model (Altúzar & Valenzuela, 1991). Isothermal experiments can provide information about the n value for different crystallisation stages, where conditions such as surface crystallisation, nucleation by ageing, etc., can be observed (Calka & Radlinski, 1988).

A typical crystallisation process in binary alloys can be described, as temperature increases, as follows. In hypoeutectic compositions (where the transition-metal content is greater than the corresponding eutectic composition for a given material), the transition metal crystallises first, until the amorphous phase reaches the eutectic composition, at which both the transition metal and the intermetallic crystallise. In hypereutectic alloys (transition-metal-poor), both phases crystallise simultaneously. Some ternary alloys can also be included in this schematic model, such as Fe–B–Si (Ramanan & Fish, 1982) and Ni–B–Si (Baró, Clavaguera and Suriñach, 1988). The crystallised phases are α-(Si–Fe) and Fe_3B for the former system, and γ-Ni and Ni_3B for the latter. Transmission electron microscopy can be used to study *in-situ* crystallisation phenomena, identify (Valenzuela *et al.*, 1982) and evaluate the crystallised fractions (Blanke-Bewersdorff & Köster, 1988).

Amorphous systems present certain phenomena not observed in crystalline materials. One such phenomenon is *structural relaxation*. This refers to changes in macroscopic properties when an amorphous material is thermally annealed; results obtained are consistent with the idea that, for each temperature, there is a preferred amorphous arrangement. This arrangement is not attained instantaneously, and there exists well-defined kinetics for relaxation towards this preferred state. This relaxation occurs at temperatures well below crystallisation and is clearly a different phenomenon to crystallisation. To complicate matters further, some of the changes are irreversible (e.g., changes in density, brittleness), while others are reversible (e.g., Curie temperature, permeability); the latter are reversible in the sense that two well-defined values can be observed when the temperature is cycled between two points. Generally, the irreversible relaxation takes place during the first high-temperature annealing, and the reversible relaxation processes occur on subsequent annealings. Changes in properties, ΔP, are typically represented by a

Soft magnetic materials

log(t) kinetics:

$$\Delta P = A \log (t) \tag{6.10}$$

where A is constant. A useful concept related to relaxation phenomena is that of the fictive temperature, T_f (Jagielinski & Egami, 1985). T_f is the temperature at which a specific short-range order configuration is in metastable equilibrium. For as-quenched samples, T_f is close to the casting temperature. Annealing for a long time at lower temperatures tends to move T_f to the annealing temperature; when the annealing temperature and T_f are comparable (ie., the sample has reached a new metastable configuration in equilibrium), no further change in properties is observed. By changing the annealing temperature, a new relaxation is driven.

By applying several experimental techniques to a particularly stable amorphous alloy (Ström-Olsen *et al.*, 1991), it was shown that irreversible relaxation is related to transition-metal atom movements leading to densification (the measured decrease in interatomic spacing was ~ 0.0026 Å); subsequent annealings at intermediate temperatures result in shear motion of transition-metal atom movements, which do not affect density; finally, at low temperatures (below 500 K), only metalloid movements are detected.

Saturation magnetisation, M_s, is always smaller in an amorphous phase than in its crystalline counterpart, but the effect of disorder on M_s is small; a substantial part of the magnetic moment reduction can be attributed to the metalloids. In the sequence: B, C, Si, Ge, P, the reduction in M_s is lowest for B and highest for P. Since binary alloys containing two transition metals are unstable, the effect of transition-metal alloying on M_s has been studied in ternary systems such as $Fe_{80-x}Co_xB_{20}$ and $Fe_{80-x}Ni_xB_{20}$. The addition of Ni to the latter results in a decrease of M_s; in the former, saturation magnetisation shows a small maximum for $x = 10$. Further addition of Co leads to a decrease in M_s, Fig. 6.19. The *Curie temperature* in amorphous materials is considerably lower than in their crystalline counterparts. As in the case of M_s, the main contributions to this reduction are disorder and alloying, but no theory has been proposed to quantify these contributions. Reductions in T_C of 12% ($Fe_{75}P_6B_{19}$), 10% (Fe_3B) and 20% (Fe_3P), as compared to the Curie temperature of the corresponding crystalline alloys, have been observed. Data on M_s and T_C for a number of alloys are given in Table 6.4.

The most important impact of disorder, as far as soft magnetic

Table 6.4. *Saturation magnetisation and Curie temperatures of some soft amorphous materials.*

Composition	M_s (kA/m)	T_C (K)
$Fe_{20}B_{20}$	1257	651
$Co_{80}B_{20}$	915	765
$Ni_{85}P_{15}$		190
$Fe_{40}Co_{40}B_{20}$	1193	>800
$Fe_{40}Ni_{40}B_{20}$	795	692
$Fe_{59}Ge_{41}$	1018	

Adapted from Luborsky (1980).

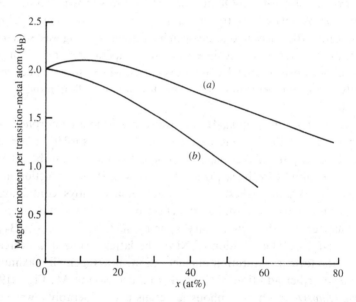

Fig. 6.19. Variations in magnetic moment per transition atom by alloying in amorphous alloys: (*a*) $Fe_{80-x}Co_xB_{20}$ and (*b*) $Fe_{80-x}Ni_xB_{20}$. (Data from O'Handley *et al.*, 1976.)

properties are concerned, is on *anisotropy*. A truly amorphous arrangement leads to a random local anisotropy; for each atom pair, magnetic moments are coupled by exchange but the easy axis and anisotropy intensity vary from one pair to the next. If short-range order existed, the easy axis would vary from one short-range group to the next. On the macroscopic scale, this is equivalent to the absence of anisotropy. An exchange-coupled material with zero anisotropy would have infinite permeability.

Experimentally, however, ribbons obtained by the melt-spinning technique show coercive fields in the range ~ 5–10 A/m. The reason for these coercivities is that, as a result of inhomogeneities produced by the severe quenching process, as-cast samples retain residual stresses. Domain structures in as-cast materials are extremely complex. Residual stresses can be relieved by appropriate thermal annealing at a temperature high enough to permit some atomic rearrangement, but low enough to prevent crystallisation; usually, this is done at ~ 50 K below the onset of crystallisation at low heating rates. In most cases, however, an induced anisotropy is retained with the easy access in the ribbon plane.

Amorphous materials are sensitive to *directional order*, in a similar manner to crystalline materials (Fujimori, Yoshimoto & Masumoto, 1981). Field annealing, i.e., thermal treatment at $T < T_C$ in a field, results in a material with an easy axis parallel to the field. A coercive field ~ 0.5 A/m and maximum permeability of 2×10^6 have been observed in annealed amorphous samples (Masumoto *et al.*, 1977). Hysteresis loops can be substantially modified by field annealing, Fig. 6.20. In *longitudinal* field annealing, domains become large and are oriented parallel to the ribbon axis. A small field can unpin and displace the domain walls, leading to huge magnetisation variations. A *transversal* field annealing results in

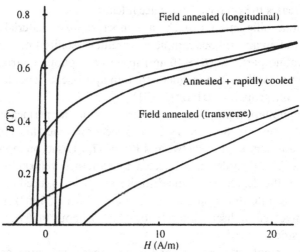

Fig. 6.20. Magnetic hysteresis loops of $Fe_{39}Ni_{39}Mo_4Si_6B_{12}$ after various thermal treatments. (After Warlimont & Boll, 1982.).

domains and walls perpendicular to the ribbon axis; magnetisation takes place mainly by spin rotation, since wall displacements have no effect on the magnetisation parallel to the applied field. A zero-field annealing can also lead to an improvement in magnetic properties, when carried out at $T > T_C$; for $T < T_C$, it becomes an *ageing* process, resulting in a degradation of permeability by domain wall pinning (see Section 6.1). The frequency response of pinned domain walls can also be widely varied by annealings; the relaxation frequency (the maximum frequency at which domain walls are able to follow the excitation field), ω_x, decreases as permeability increases. As a result of ageing, pinning determines the size of the bowing wall and therefore the permeability value, but the maximum 'vibration' frequency of the wall varies as the inverse of wall size (Valenzuela & Irvine, 1992). Experimental results on Co-based amorphous ribbons can be fitted with a phenomenological relationship $\mu_i^\alpha \omega_x = $ constant, with $\alpha \approx 1$–1.5 (Valenzuela & Irvine, 1993).

Stress-annealing, i.e., thermal treatment under a tensile load, can also lead to induced anisotropy. The phenomena are more complex than for field annealing; at low temperatures, the easy direction is parallel to the applied stress; at high temperatures, it becomes orthogonal (Gibbs, 1990). A third method for inducing anisotropy in amorphous alloys is surface crystallisation. Since the corresponding crystalline phase usually has higher density, the crystalline layer exerts a compressive stress on the amorphous matrix. A shortcoming of this method is that crystallisation usually increases the coercive field, since it leads to increased domain-wall pinning. In a different approach, a nanocrystalline material can be obtained from an amorphous sample by careful crystallisation. This gives rise to ultrafine particles ($D \approx 10$ nm) in an amorphous matrix. Anisotropy becomes random and magnetostriction approaches zero, leading to favourable soft properties (Herzer, 1991).

Disaccommodation phenomena (a decrease in initial permeability as a function of time, see Section 4.4.2) also occur in metallic glasses, at practically any temperature between 4 K and T_C. Disaccommodation is related to atomic disorder; it is affected by structural relaxation and by differences in the degree of atomic disorder and is introduced by varying the quenching rate during preparation (Allia & Vinai, 1987). Diffusion processes involved in disaccommodation of amorphous materials are not fully understood.

From the point of view of *applications*, there are three groups of amorphous ribbons: Fe-based alloys, which are inexpensive with high

Table 6.5. *Applications of amorphous alloys.*

Power distribution transformers
Switched-mode power supplies (SMPSs)
400 Hz transformers
Electromagnetic shielding
Magnetomechanical transducers
Security systems

saturation induction; Co-based alloys with low saturation induction, extremely low magnetostriction and the highest permeability; Fe–Ni alloys, with intermediate properties. Some of their most important applications are listed in Table 6.5. The higher resistivity of amorphous alloys has attracted much interest for applications in power distribution transformers, since a significant reduction in losses could be expected. However, two factors involving significant changes in transformer design have delayed their use: one is that saturation induction in amorphous alloys is considerably lower than in grain-oriented Si–Fe; the other is that they have a limited width (~ 17 cm). These factors are not important for small, 400 Hz transformers for ships and aircraft. Due to their resistivity, amorphous cores permit a substantial reduction in volume and weight in this kind of device.

In SMPSs (see Section 5.3), a high-frequency transformer is essential. For working frequencies up to ~ 100 kHz, amorphous materials have losses comparable to those of ferrites, but possess higher saturation induction.

Amorphous ribbons possess not only good magnetic properties, but also outstanding mechanical properties. For example, an elastic strain of 1% is feasible in amorphous alloys; in crystalline samples, 0.1% is hardly possible (Hinz & Voigt, 1989). Zero-magnetostriction alloys (Co-based) conserve their elevated permeability in spite of bending, and can be used in flexible magnetic shielding. Magnetomechanical transducers are based on the change in magnetic properties as a result of the application of a mechanical stress (the 'Villari' principle). Security systems make use of the fact that, when a soft ferromagnet is excited by an ac field of a given frequency, it generates odd harmonics of the original frequency. 'Perminvar' alloys, which are self-magnetic annealed materials with constant permeability at low fields, can generate even and odd harmonics.

In amorphous, perminvar-like ribbons (Co-based, aged at $T < T_C$), an additional non-linear term proportional to H^3 appears. Identification can be made by means of the signal profile for a given excitation frequency (Hasegawa, 1991). Small ribbons on books, for example, can be used as an effective security system.

Amorphous alloys containing a *rare earth–transition metal* combination offer the possibility of studying some of the most interesting fundamental aspects of magnetism: magnetic interactions, random anisotropy, non-collinear structures, dilution effects, etc.; these are beyond the scope of this overview (Cornelison & Sellmyer, 1984; Krishnan, Lassri & Rougier, 1987). Some of these materials are used as thin films in magnetooptic applications; a short review is given in Section 6.4.

Spin glasses are metallic alloys of a diluted 3d atom (Fe, Co, Ni, Mn) in a non-magnetic matrix (Mydosh & Nieuwenhuys, 1980; Maletta & Zinn, 1989). At low temperatures and low dilutions, a localised anti-ferromagnetic interaction of the isolated magnetic moment with the conduction electrons can take place. This is the so-called *Kondo* effect. As the impurity concentration increases, there may exist a temperature, T_f, below which impurity spins become 'frozen' in random directions. This spin arrangement is known as *spin glass*. The magnetic interaction is propagated over long distances by conduction electrons. If this interaction promotes a parallel arrangement, a non-uniform ferromagnetic order can be established even at low impurity concentrations. Some systems exhibit spin-glass behaviour at low concentrations and non-uniform ferromagnetic order at high concentrations, separated by a *percolation* limit. This percolation limit corresponds to the concentration at which each magnetic site has, at least, one magnetic nearest neighbour. Then, a direct interaction can be uninterruptedly established through the whole sample. In the case of an antiferromagnetic interaction, *frustration* occurs when a given site is subjected to the interaction of the two neighbouring atoms with opposite orientation, Fig. 6.21. Under these

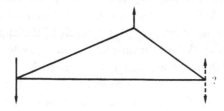

Fig. 6.21. 'Frustration' in an amorphous antiferromagnet (schematic).

conditions, no long-range order can exist. On cooling, this material becomes a spin glass. Examples of spin glass systems are Mn in Cu, Fe in Au, Mn in Ag, Co in Pt. Spin glasses cannot be identified by a single experiment; they are recognised by a combination of characteristics (Gignoux, 1992).

6.2 Hard magnetic materials

The area of hard magnets has exhibited an impressive development since the late 1960s, with new materials with exciting properties. The basic parameter to characterise the quality of a permanent magnet, the energy product $(BH)_{max}$ (see Section 4.5.2), has been increased by more than two orders of magnitude since the introduction of the first 'hard steels', Fig. 6.22. Devices based on permanent magnets are increasingly smaller and more efficient. New magnets with a large energy product can be used

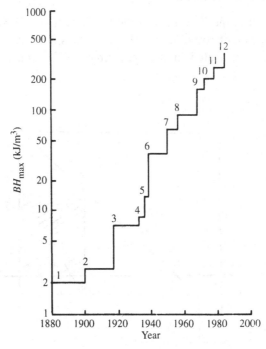

Fig. 6.22. Increase in the energy product of permanent magnets between 1880 and 1980: 1, 2, 3, steels; 4, Fe–Ni alloy; 5, 6, 7, 8, alnicos; 9, $SmCo_5$; 10, $(SmPr)Co_5$; 11, $Sm_2(Co_{0.85}Fe_{0.11}Mn_{0.04})_{17}$; and 12, $Nd_2Fe_{14}B$, (From Jiles, 1991.)

257

to produce magnetic fields instead of coils with electrical currents. Permanent magnets are more efficient than the latter, since they have no ohmic losses. For materials with a high coercive field, there is a substantial difference between the $B-H$ and the $M-H$ plots, Fig. 6.23, since B depends also on the applied field. To distinguish between them, the value of the coercive field measured from $B-H$ plots is represented as $_BH_c$ and the one determined from magnetisation measurements as $_iH_c$; $_iH_c$ is *intrinsic*, since it does not depend on the applied field.

As discussed in Section 4.5, the desired properties in materials for permanent magnet applications are essentially a high remanent induction, B_r, a large coercive field (either $_BH_c$ or $_iH_c$) and a large energy product $(BH)_{max}$. A brief discussion of the various mechanisms leading to increased coercivity is presented first. Since most of the new hard magnets are intermetallic compounds, a short account of their anisotropies is included.

6.2.1 Coercivity mechanisms

The magnetisation mechanism with the highest coercive force is domain rotation (or coherent rotation), in which all the spins within the sample are collectively reoriented in the field direction. In uniaxial materials,

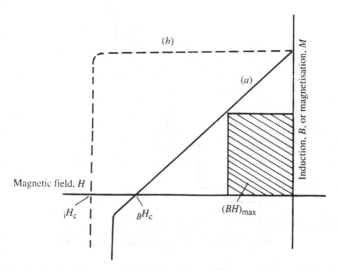

Fig. 6.23. Comparison between (a) induction, B, and (b) magnetisation, M, curves, as a function of applied field to illustrate the difference between coercivity, $_BH_c$, and intrinsic coercive field, $_iH_c$.

this occurs when the applied field attains the anisotropy field value, $H_K = 2K_1/M_s$, where K_1 is the magnetocrystalline anisotropy constant and M_s the saturation magnetisation. Both parameters depend almost entirely on the material composition, and are considered as *intrinsic*, together with Curie temperature, T_C. Experimentally, however, the coercive field value for a given composition can vary widely, but its maximum observed value is usually about $H_K/10$. The coercive field depends on anisotropy, but is strongly affected by defects, grain size, grain shape and distribution, etc.

The difference between the experimental coercive field and the anisotropy field (as determined from the magnetocrystalline anisotropy constant) is explained by the presence of domain walls. As discussed in various sections in this book, the propagation of domain walls leads to huge changes in magnetisation, even at extremely low applied fields. An effective increase in coercivity can be obtained by domain wall *pinning*; i.e., by impeding wall displacements. Domain wall pinning can be developed by introducing a number of inhomogeneities into the material. Dissolving 1 wt% C in Fe increases the coercive field by up to ~ 4 kA/m in low C steels, the first hard magnets. This effect is attributed to the creation of a stress field around each interstitial C, which impedes domain wall motion. In polycrystalline materials, grain boundaries act naturally as pinning sites for domain walls. Their effects can be enhanced by a non-magnetic second phase at the grain boundaries; this mechanism is important in a number of hard materials. Another approach consists of adding a second phase to stabilise the domain walls. A fine, dispersed second phase is more effective than one in the form of coarse, spherical precipitates, because in the latter, domain walls can easily move between large obstacles. In some alloys such as the Alnicos, the equilibrium structure is formed by two different phases; the coercivity mechanism involves the controlled transformation from the single phase at high temperature to the two-phase structure at lower temperature. This case is discussed later in this section.

In hard ferrites, a technique to avoid wall movement is simply to eliminate domain walls by reducing the particle size below the critical, single-domain value. In high anisotropy intermetallics such as $Nd_2Fe_{14}B$, even for grain sizes in the critical range, no clear single-domain behaviour has been observed. There exists a different mechanism that can explain the low experimental coercive fields, especially in high anisotropy materials: this is domain wall *nucleation* (Zijlstra, 1980). In this mechanism, a domain can be nucleated and grown in a direction favourable to lower the total energy of the system. The nucleation process can take place near defects, where variations in *local* values of anisotropy field facilitate a magnetisation reversal. A nucleation field, H_n, characterises this process. Surface defects are also important; the smoothness of the external surface can lead to substantial differences in the hysteresis loop shape (McCurrie & Willmore, 1979).

Magnetisation reversal in most intermetallic hard magnets seems to be nucleation-controlled (Buschow, 1986). Domain wall pinning, however, is important to conserve a high coercivity. In a magnetically saturated material with $H_n > H_p$ (where H_p is the pinning field), for instance, magnetisation reversal is initiated once the applied field reaches the nucleation field, H_n. In principle, it is enough to nucleate *one* domain wall to produce an abrupt reversal of magnetisation; a single domain wall can propagate through the whole sample since the applied field is larger than the value required for unpinning. In the case $H_p > H_n$, domains are nucleated as $H = H_n$ and a small decrease in magnetisation can be expected as a result of the nucleated volume. However, domain walls remain pinned and no reversal of magnetisation is observed until $H = H_p$. Schematic demagnetisation curves for these two cases appear in Fig. 6.24. Experimental hysteresis loops include the effects of reversible domain wall bowing, which leads to considerable rounding of the loops and makes it difficult to distinguish between the two mechanisms (Cook & Rossiter, 1989).

6.2.2   Alnico alloys

Alloys of Fe, Ni, Co and Al (among the major components) are important not only because they constitute $\sim 6\%$ of the total world production of

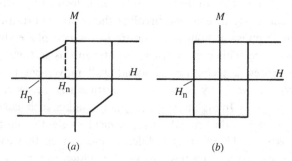

Fig. 6.24. Schematic hysteresis loops in a material where magnetisation reversal occurs by: (*a*) domain wall unpinning, and (*b*) by domain nucleation. H_n and H_p are the nucleation and pinning fields, respectively.

Table 6.6. *Typical properties of some Alnicos.*

Material	B_r (T)	$_BH_c$ (kA/m)	$(BH)_{max}$ (kJ/m^3)
Alnico 8 (field treated)	0.8–0.9	110–140	40–45
Alnico 8 (extra high H_c)	0.76	150–170	44–48
Alnico FDG (oriented grain)	1.3–1.4	56–62	56–64
Alnico 9 (oriented grain)	1.0	110–140	60–75

Adapted from McCurrie (1982).

hard materials, but they are also interesting due to their coercivity mechanism. At room temperature, an Alnico sample shows an extremely fine two-phase microstructure with interlocking, rod-like precipitates of a Fe–Co-rich phase in a Ni–Al-rich matrix. The Fe–Co phase is strongly ferromagnetic, while the Ni–Al matrix is weakly magnetic. The precipitates can be as small as $300 \times 300 \times 1200$ Å (Cullity, 1972). The high coercive field of Alnicos is produced by shape anisotropy of the single-domain particles of the Fe–Co phase. Typical properties for the most highly developed commercial Alnicos are listed in Table 6.6; a detailed description of their preparation processes has been given by McCurrie (1982).

The Alnico microstructures are prepared by a process which involves *spinodal* decomposition. In this phase transformation, the high-temperature phase decomposes into two phases, usually known as α_1 and α_2, Fig. 6.25. A spinodal curve inside the solvus curve separates the regions where either spinodal decomposition (compositions and temperatures inside the spinodal curve) or normal, nucleation and growth transformation (between solvus and spinodal) occur. Spinodal decomposition occurs by periodic composition fluctuations (Burke, 1965); as transformation proceeds, composition fluctuations increase (α_1 becomes richer in A and α_2 in B, for instance), but the spatial periodicity is conserved.

The main operations involved in preparation of an Alnico hard magnet are as follows. The high temperature phase is treated at 1520 K to obtain an homogeneous solid solution; it is cooled at a controlled cooling rate, in the region of 50 K/min, to 700 K. During cooling, spinodal decomposition occurs; the resulting microstructure depends on the cooling rate. The alloy is then reheated ('temper' treatment) to 870 K and maintained for a few hours. This last process results in an increased magnetisation of the Fe–Co phase, presumably by diffusion of additional Fe and Co from the matrix to the Fe–Co phase.

There are two important improvements in the Alnico preparation method. The first is to perform the cooling operation in a magnetic field. During spinodal decomposition, the Fe–Co phase particles grow with their axes parallel to the applied field, leading to an easy direction for the whole sample. The second is that, under certain conditions, Alnico alloys can be directionally solidified, which results in a strong preferential orientation. This can be done by imposing a directional heat flow during alloy casting, Fig. 6.26. If heat is extracted preferentially from the bottom of the mould, the first solidified grains appear at the bottom. As solidification proceeds, columnar grain growth takes place, resulting in a strong preferential orientation. The magnetic behaviour of such a sample is significantly improved, Fig. 6.27, as compared with that of other Alnicos.

Because of their high Curie temperature (~ 1070 K) and thermal stability, Alnico materials are used in high-temperature applications and measuring instruments. They are also used in loudspeakers, but strong competition from cheaper hexagonal ferrites is decreasing their share of this market.

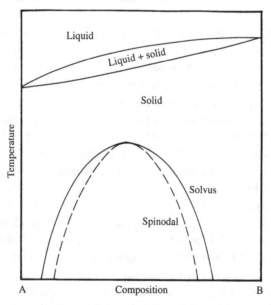

Fig. 6.25. Phase diagram of a spinodal decomposition (schematic).

Table 6.7. *Magnetic properties of SmCo₅.*

$(BH)_{max} = 210\,\text{kJ/m}^3$
$_iH_c = 4.4\,\text{MA/m}$
$T_C = 995\,\text{K}$
$K_1 = (11-20) \times 10^6\,\text{K/m}^3$

From Strnat (1988).

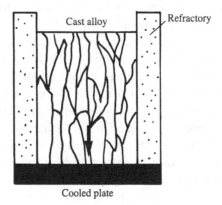

Fig. 6.26. Directional heat flow during alloy casting to induce columnar grain growth (schematic).

6.2.3 Sm–Co magnets

At the end of the 1960s, a new material with exceptional hard magnetic properties (Table 6.7) was prepared: samarium–cobalt, SmCo₅. Its most outstanding property is its magnetocrystalline anisotropy; reported values are in the range $(11-20) \times 10^6\,\text{J/m}^3$ (by comparison, Ba ferrite has an anisotropy of $0.33 \times 10^6\,\text{J/m}^3$, Table 4.15). Rather than being a serendipitous discovery, the successful synthesis of SmCo₅ was the result of a systematic effort (Strnat, 1988).

The combination of a 3d transition metal (Cr, Mn, Fe, Co, Ni, Cu, Zn) and a 4f rare earth (Ce, Pr, Nd, Pm, Sm, Eu, Ge, Tb, Dy, Ho, Er, Tm, Yb or Lu; occupancy of the orbital 4f goes from $4f^1$ for Ce, to $4f^{14}$ for Lu) usually leads to well-defined compounds, or intermetallics, which, in contrast to solid solutions, have a constant, integral ratio of atoms. There is an extremely wide variety of crystal structures, and even the binary phase diagrams are complex. Certain 'families' of compounds exist with the same crystal structure. These include combinations such as R_3M, R_3M_2, RM_5, R_2M_{17}, where R and M represent rare earth and 3d transition metals, respectively. A systematic study therefore gives insight into the general trends linking structure to intrinsic properties.

The RM_5 family for M = Co, Ni, Cu, Zn, has the $CaCu_5$-type hexagonal structure, as shown in Fig. 6.28. The band structure, formed by contributions from the 3d, M, and 4f, R, bands, changes in such a way that for $SmCo_5$:

(1) The magnetic moment at M sites is maximum for Co;

(2) Co–Co nearest neighbours are found in the *c*-axis direction, leading to a strong ferromagnetic exchange interaction, characterised by a high Curie temperature ($\sim$950 K);

(3) a strong Sm–Co ferromagnetic exchange interaction occurs, one order of magnitude lower than that of Co–Co exchange;

(4) the Sm–Sm interaction is antiferromagnetic, but one order of magnitude smaller than the Sm–Co exchange;

(5) the *c*-axis is the easy direction both for Co and Sm; magnetocrystalline anisotropy is extremely high due to an additional contribution from the hexagonal structure.

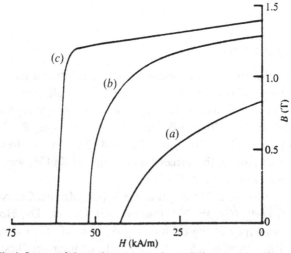

Fig. 6.27. The influence of the various preparation techniques on the demagnetisation curves of three representative Alnicos: (*a*) isotropic; (*b*) field cooled; and (*c*) grain-oriented and field cooled. (Adapted from Gould, 1959.)

Hard magnetic materials

$SmCo_5$ is an exceptionally hard magnetic material because of the combination of all the above factors (Gignoux, 1992).

Intense research on similar materials led to compounds of the type R_2M_{17} (or '2–17', to distinguish them from the '1–5' compound) where M is the combination of several metals; an example is $Sm(Co_{0.67}Fe_{0.23}Cu_{0.08}Zr_{0.02})_{8.35}$ (Ray *et al.*, 1987). Compounds with the 2–17 stoichiometry can have either a hexagonal or a rhombohedral unit cell, Fig. 6.29. The high (intrinsic)

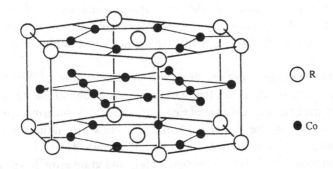

Fig. 6.28. The hexagonal unit cell of $SmCo_5$ ($CaCu_5$ type).

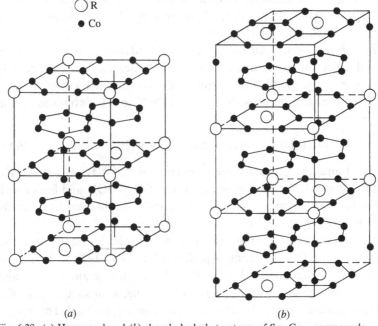

(a)   (b)

Fig. 6.29. (*a*) Hexagonal and (*b*) rhombohedral structures of Sm_2Co_{17} compounds.

265

coercive field obtained in this compound, 1.8 MA/m, was the result of a carefully prepared microstructure. The grains consisted of small cells (100–200 nm) with the 2–17 overall composition, enveloped by a thin layer (5–20 nm) of a 1–5 phase. There were also reported to be very thin platelets of a Zr-rich phase, crossing many cells and boundaries; all three phases were crystallographically coherent. The high coercivity was attributed to domain wall pinning in this microstructure (Hadjipanayis, Yadlowski & Wollins, 1982). Zr additions were needed for the platelet phase; the precipitation of the enveloping phase occurred with Cu additions; finally, Fe addition led to an increase in the remanent induction value. A review of the details of many other substitutions is given by Strnat (1988).

Permanent magnets are prepared by *powder metallurgy*, i.e., by sintering metallic powders. There are two basic methods to obtain the powder: by fusion of the corresponding metal elements, and by calciothermic reduction of the oxides. In the *fusion* method, the metals are induction-melted, arc-melted or simply melted in an electric furnace (1500–1700 K) and chill-cast (water cooled) to avoid segregation and grain growth. The ingots are annealed at ~ 1420 K for 1–5, and 1370 K for 2–17 compounds. After cooling, ingots are crushed and finely ground to a particle size of ~ 5 μm. Powders are extremely reactive; storage in a dry Ar or N_2 atmosphere is needed to avoid oxidation.

Calciothermic reduction has an economic advantage over the fusion method, since the rare earth is used as an oxide, which is considerably cheaper than the metal. The transition metals are used as fine metallic powders. The reaction is carried out at 1420 K in a H_2 atmosphere and can be written:

$$Sm_2O_3 + 10Co + 3Ca \rightarrow 2SmCo_5 + 3CaO \qquad (6.11)$$

The CaO and any Ca excess are eliminated as $Ca(OH)_2$ in wet H_2; the product is rinsed, first with water, then with CH_3COOH and finally with C_2H_5OH. After drying in vacuum, the powder is obtained by crushing and grinding.

To obtain the final permanent magnet, the powder is oriented in a field prior to compaction. Sintering is performed at 1400–1420 K in an inert atmosphere (vacuum, Ar, He or H_2). To control grain growth and accelerate densification by *liquid-phase* sintering, a small quantity of a binary composition near the Sm_2Co_7 stoichiometry is added. The highest coercivities are obtained by annealing at 1159 K and rapidly cooling below 600 K. Sintering is far more complicated in 2–17 compounds, especially if the complex, three-phase microstructure (Cu-hardened) is to be obtained (Strnat, 1988).

Bonded magnets have many applications; a wide variety of shapes can be obtained by compaction, extrusion, injection, etc. One of the basic problems, however, is the high chemical reactivity of the alloys. Instead of polymers, a metallic alloy, such as Sn–Pb solder, or Cu can be useful as a matrix material. Cu-hardened magnets are more stable against oxidation.

Because of their high energy product, most *applications* of $SmCo_5$ and Sm_2Co_{17} magnets are related to the miniaturisation of devices. Also, some small motors (and generators, actuators) which were traditionally operated with electromagnets, have been redesigned to use permanent magnets. A serious shortcoming of this group of magnets is the high price of Sm and Co. In the high energy-product range of materials, $Nd_2Fe_{14}B$ magnets (see next section) are chaper and in many respects perform better; they substitute for $SmCo_5$ in most applications. $SmCo_5$ magnets are therefore expected to remain in use for specialised devices, where the final price does not depend strongly on the magnetic cost. This is the case for small dc motors in portable devices such as cassette players and video cameras.

6.2.4   Nd–Fe–B magnets

The high cost of $SmCo_5$ hard materials promoted the search for alternative cheaper high-energy product magnets. Many binary RM alloys were investigated, especially those containing Fe, but the few stable intermetallics showed a Curie temperature which was too low, and had no uniaxial anisotropy. A ternary compound in the system Nd–Fe–B with impressive properties was found simultaneously in 1984 by two different preparation methods: induction melting (Sagawa *et al.*, 1984), and melt-spinning (Croat *et al.*, 1984). Values of coercive field, $_iH_c = 960$ kA/m, remanent induction, $B_r = 1.23$ T, energy product $(BH)_{max} = 290$ kJ/m^3 and Curie temperature, $T_C = 585$ K, were reported (Sagawa *et al.*, 1984). The composition of the phase was found to be $Nd_2Fe_{14}B$ with a tetragonal structure (Herbst *et al.*, 1984), Fig. 6.30.

The experimental value of the saturation magnetisation at room temperature for practically the whole family of compounds $R_2Fe_{14}B$ (R = rare earth) can be explained on the basis of a ferromagnetic coupling between both Fe sublattices and the rare-earth sublattice. Some compounds, however, show spin reorientation phenomena at low temperatures. For R = Er, Tm or Yb, the easy magnetisation direction deviates from the *c*-axis as temperature decreases; at 4.2 K, the magnetisation is found in the basal plane. Spin reorientation in these compounds seems to be associated with an antiferromagnetic coupling between the Fe sublattices and the rare-earth sublattice. A detailed account of these phenomena is found in Buschow (1988).

The potential economic and technological impact of these materials prompted the organisation of a collective effort by the Commission of the European Communities in 1985, known as 'Concerted European Action in Magnets', CEAM. Many discoveries have resulted from this effort.

$Nd_2Fe_{14}B$ permanent magnets can be prepared by induction melting, followed by liquid-phase sintering of powder, in a similar way to $SmCo_5$ materials. On cooling the melt, the first solid formed is γ-Fe. To prevent

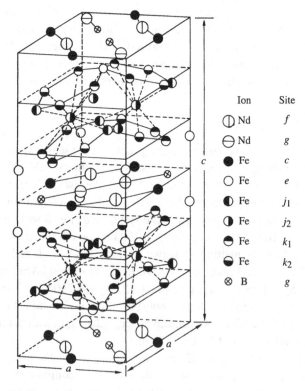

Ion	Site
Nd	f
Nd	g
Fe	c
Fe	e
Fe	j_1
Fe	j_2
Fe	k_1
Fe	k_2
B	g

Fig. 6.30. The unit cell of $Nd_2Fe_{14}B$, showing the different Fe sites. (Adapted from Croat *et al.*, 1984)

the separation of a significant quantity of Fe, the sample is cooled rapidly. Since the powders are pyrophoric (flammable in contact with air), extreme care has to be taken during all operations, but most critically during milling. Milling is usually done in an organic solvent, such as isopropanol or toluene, to avoid contact with the air.

A particular phenomenon, known as *hydrogen decrepitation*, can be used to prepare high-quality magnets (McGuiness *et al.*, 1986). In this method, bulk ingots of the initial alloy are subjected to a H_2 atmosphere; in just a few minutes, a strong attack transforms the solid into a fine powder. An advantage of this method is that decrepitation takes place along well-defined crystallographic planes, yielding single-crystal particles with minimal mechanical damage.

The calciothermic reduction method can also be used to prepare the $Nd_2Fe_{14}B$ sintering powder. The reaction in this case is:

$$2Nd_2O_3 + B_2O_3 + 28Fe + 9Ca \rightarrow 2Nd_2Fe_{14}B + 9CaO \qquad (6.12)$$

As in the synthesis of $SmCo_5$, excess Ca is eliminated; after rinsing, crushing, milling and drying, the powder is ready for compaction and sintering.

Liquid-phase sintering is carried out at 1380 K in vacuum or Ar; after quenching to room temperature, an annealing at 880 K results in a substantial increase in coercivity. The initial composition is close to $Nd_{15}Fe_{77}B_8$, which is richer in Nd and B than that of $Nd_2Fe_{14}B$; this leads to a small amount of liquid at the sintering temperature. After sintering, the sample is quenched to room temperature. Annealing at 880 K doubles the coercive field; presumably, this treatment has an effect on the grain boundary phases, which are thought to be a mixture of $NdFe_4B_4$ and a Nd-rich phase.

Use of the *melt-spinning* method (see Section 6.1.2) to produce the initial alloy involves preparation of an amorphous intermediate phase (Ormerod, 1985). For high tangential speeds of the wheel, > 30 m/s (high quenching rates), a completely amorphous phase was obtained with a small coercive field. For low tangential speeds (lower than 15 m/s), relatively large crystalline grains were observed, also with small coercivity. At the intermediate speeds (16–20 m/s), coercive fields in the range 1100 kA/m were measured; samples contained spherical grains 20–100 nm in diameter. The best fabrication procedure was to obtain 'overquenched' ribbons and adjust the coercive field by means of a subsequent careful heat treatment. Bonded magnets could then be prepared by milling the ribbons and mixing with an appropriate polymer. A shortcoming of this method is that samples are isotropic; coercivity is relatively small. Full densification can be obtained by hot pressing, with reported $_iH_c$ values in the region of 115 kJ/m^3 (Gwan *et al.*, 1987).

The plastic deformation of milled ribbons leads to a preferential grain orientation in melt-spun $Nd_2Fe_{14}B$ magnets (Lee, Brewer & Schaffel, 1985). When such a sample was uniaxially pressed at high temperature, plastic flow (for deformations of $\sim 50\%$) resulted in a deformation of the spherical grains into platelets oriented parallel to the flow direction. The easy axis (*c*-axis) became parallel to the pressing axis, and a $(BH)_{max}$ value of 320 kJ/m was observed. To obtain this deformation, known as *die-upsetting*, a die with a cavity of larger diameter was used. The mechanism of preferential orientation formation is related to an anisotropic growth of $Nd_2Fe_{14}B$ grains surrounded by a film of a Nd-rich grain boundary phase (Mishra, Chu & Rabenberg, 1990).

A further fabrication method, known as *mechanical alloying* leads to an amorphous material by ballmilling of the elements in a closed container, followed by heat treatment to homogenise and crystallise the compound (Cahn, 1990). Usually, this method results in isotropic magnets.

The main features of the most important fabrication routes for permanent magnets are shown in Fig. 6.31.

The *applications* of permanent magnets with high energy products are in motors and generators for use in domestic appliances, computers, dc motors in cars, industrial servo-machines and medical instruments. A significant reduction in size can be obtained with these high-performance materials; in Fig. 6.32, a loudspeadker fabricated with $Nd_2Fe_{14}B$ is compared with the equivalent device containing instead other permanent magnet materials. In many cases, however, ferrite, because of its low cost, still dominates the market.

6.2.5 Nitrides and carbides

The relatively low Curie temperature of $Nd_2Fe_{14}B$ and other shortcomings (low corrosion resistance, relatively high price, etc.) has led to the investigation of other intermetallic systems. The main families are $R_2Fe_{14}C$ and $R_2Co_{14}B$ (isomorphous with $Nd_2Fe_{14}B$), $R_2Fe_{17}C_x$ (hexagonal or rhombohedral) (Buschow, 1991) and $R_2Fe_{17}N_x$ (Coey & Sun, 1990), where R is a rare earth. In these four systems, the role of the metalloids is the

Hard magnetic materials

same: they enter interstitially, stabilise the anisotropic structure, expand the unit cell (by $\sim 7\%$) and lead to an increase in Curie temperature and saturation magnetisation. The last two characteristics are not purely volumetric effects; they seem to be related to substantial changes in band structure associated with the presence of the interstitial atom (Coey &

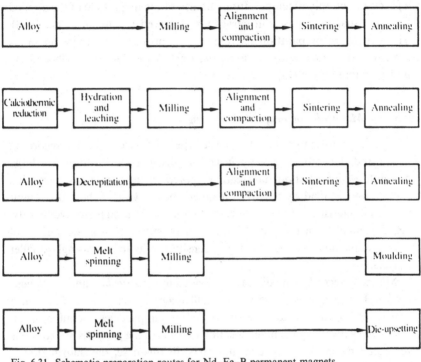

Fig. 6.31. Schematic preparation routes for Nd–Fe–B permanent magnets.

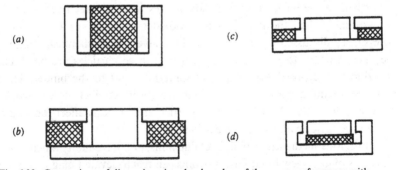

Fig. 6.32. Comparison of dimensions in a loudspeaker of the same performance with different permanent magnets: (a) alnico; (b) ferrite; (c) $SmCo_5$; and (d) $Nd_2Fe_{14}B$. (After Cartoceti, 1985.)

Hurley, 1992). A wide variety of magnetic properties is found; in most cases, if R has no magnetic moment or is a light rare earth, the spin arrangement is ferromagnetic; with heavy rare earths, an antiparallel coupling is established with the Fe sublattice, as in $R_2Fe_{14}B$ compounds. In many cases, the transition from one particular magnetic structure to another is a function not only of temperature, but also of the magnitude of the applied field; unusual magnetisation curves are observed (Cadogan *et al.*, 1988). The structures, preparation techniques and properties of these materials are currently under study (Zhao *et al.*, 1993; Altounian *et al.*, 1993; Gong and Hadjipanayis, 1993).

6.3   *Materials for magnetic recording*

Magnetic recording requires two basic types of device: magnetic recording *media* and magnetic recording *heads*. As discussed in Section 5.4, large amounts of information data can be stored in tapes and discs (media), written and/or read by means of magnetic heads. One of the most sought after characteristics in magnetic recording media is a high storage density, i.e., the quantity of information that can be stored in a given surface of magnetic medium. Higher recording density results in increasingly smaller devices.

Magnetic recording media can be divided into *particulate* and *continuous* media. The former is made up of small magnetic particles bonded on a plastic tape or disc; since these are single-domain particles, the information is stored by inverting the magnetisation of some of the particles. Continuous recording media are magnetic thin films; in this case, the information is 'written' by creating domains with different orientations.

Particulate media have certain advantages: they are relatively easy to prepare, and when the particulates are oxides (γ-Fe_2O_3, or Co-modified γ-Fe_2O_3, as discussed in Section 5.4), they have a high chemical stability; they are resistant to wear, not only because these particles are hard, but also due to the possibility of adding some lubricant to the binder. Their main disadvantages are that recording density is relatively low; particles are somewhat diluted, since particle–particle interactions affect the single-domain behaviour and must be avoided.

Continuous media are usually metallic films, and have many advantages: thin films of just a few hundred angstroms can reach elevated magnetisations and coercive fields (as high as 2000 kA/m, and 159 kA/m, respectively; Bate, 1980); another advantage is that, in principle, the shape of hysteresis loops can be widely varied in a controlled way. A number of problems, however, have retarded their massive application in many devices. These problems have their origin in the metallic nature of the films: they are extremely prone to corrosion; the friction caused by a mechanically hard recording head produces accelerated wear; also, a continuous film can have adhesion problems when deposited onto flexible substrates.

Particulate media of the ceramic (oxide) type have been discussed in Section 5.4; in this section, only metallic particles (Fe) are briefly described. Fe particles possess a high remanent induction in the 0.30–0.5 T range, and intrinsic coercive fields as large as 115 A/m; as a comparison, a hysteresis loop of Fe particulate medium is one order of magnitude larger than one of γ-Fe_2O_3 (Mallinson, 1987). Fe particles suitable for magnetic recording are prepared from γ-Fe_2O_3, which is easily obtained as small, elongated particles. They are reduced to metallic Fe by heat treatment at 570 K in H_2. To avoid sintering and agglomeration of particles, a surface coating is provided prior to reduction, usually by means of an aqueous solution of $SnCl_2$. The Fe particles require another coating to protect them against oxidation from air contact; with such an extended surface, reaction with air can result in a spontaneous ignition. This coating is produced by treatment in N_2, to form an outer layer of nitride.

Metallic films for magnetic recording media can be prepared by four basic methods (Bate, 1980): electromechanical deposition (or electroplating), chemical deposition, evaporation and sputtering. Most of the research in this area has been done on Co alloys obtained by sputtering; this method allows an accurate control of the thickness and composition of thin films. Co alloys provide a high remanent induction (higher than Ni and comparable to Fe alloys), and a high anisotropy (higher than both Ni and Fe alloys). Co–Cr alloys, in particular, are the most widely investigated materials for thin-film magnetic recording because of their outstanding properties.

Co–Cr alloys have an hcp structure (for Cr < 20 at%), and under certain conditions, show a strong tendency to columnar growth with the *c*-axis perpendicular to the plane of the substrate. This tendency is enhanced when the film deposition is made on a Cr underlayer and is attributed to an epitaxial growth occurring preferentially on the basal planes. The resulting columns have a diameter of typically ~1000 Å (Mallinson, 1987), which is in the range of single-domain behaviour, Fig. 6.33. Since the easy direction is the *c*-axis, the magnetisation is perpendicular to the plane of the film. This is the basis of *perpendicular*, or vertical magnetic recording, which is expected to lead to higher recording densities (Iwasaki, 1984). Columnar Co–Cr alloys have been obtained on a variety of underlayers; presumably, the stresses produced by a large lattice mismatch are relieved at the grain boundaries, leading to microvoids. These microvoids, in turn, result in an increase in coercive field.

Co–Cr alloys are extremely flexible; a thick underlayer can change the *c*-axis orientation to maintain the magnetisation in the plane of the substrate, leading to materials useful for *longitudinal* recording as well (Tsumita, 1987). Investigation of the effects of underlayers, interlayers (Sellmyer, Wang & Christner, 1990) and substrates (Kogure, Katayma & Ishii, 1990) is currently extremely active. Also, a new perpendicularly-oriented material with high saturation magnetisation and coercivity has been discovered; its composition can be written as: $Co_{69}Pt_{20}B_6O_5$ (Hayashi *et al.*, 1990). The O content was observed to have a strong influence on coercivity, but the detailed mechanism is not fully understood.

The ideal properties of magnetic recording *heads* are listed in Table 5.1. Ferrite materials for heads were discussed in Section 5.4.3; the main advantage of metallic materials over ferrites is their higher saturation magnetisation, which leads to higher flux densities. In principle, many of the soft metallic alloys discussed in Sections 6.1 and 6.2 could be used as head material. However, eddy currents strongly limit the permeability response at high frequencies.

A solution to eddy-current problem is the fabrication of heads with a combination of materials: instead of a massive metallic body, there is only a metallic thin film in the gap of a ferrite head, Fig. 6.34. The magnetic flux produced by the ferrite ring is strongly enhanced in the most important area of the head by a thin film of a metallic alloy with high saturation magnetisation. Eddy currents are therefore limited by the film

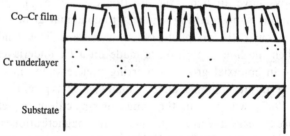

Co–Cr film

Cr underlayer

Substrate

Fig. 6.33. Schematic structure of a Co–Cr thin film for perpendicular recording.

thickness, while magnetic flux density can attain values close to those corresponding to the metallic alloy. These heads, also known as metal-in-gap heads (MIG), are usually prepared by sputtering a permalloy (Ni–Fe) in the gap; to stabilise the gap, instead of simply air, a thin film of SiO_2 glass is deposited between the two metallic films and bonded by heating. To improve the wear resistance, the permalloy films are substituted by Sendust (Al–Fe–Si), since this alloy is extremely hard.

The interfaces between the ferrite and the metallic films can affect the head performance, since they represent two additional gaps. Based on purely geometrical considerations, these effects can be greatly reduced by changing the interface angles with respect to the gap, Fig. 6.34(c)

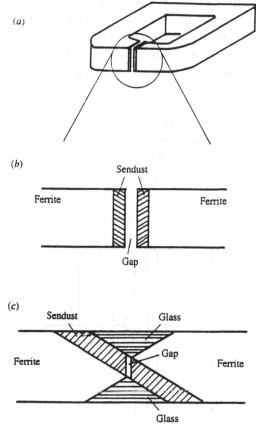

Fig. 6.34. Metal-in-gap (MIG) recording heads, made with a combination of ferrite and metallic thin films: (a) schematic representation; (b) gap enlargement; and (c) optimisation by geometric arrangement of the metallic thin film. (Adapted from Kobayashi *et al.*, 1985).

275

(Kobayashi *et al.*, 1985). A high-performance recording head design was proposed by Ruigrok, Sillen & van Rijn (1990).

Perpendicular recording heads require magnetic flux densities even higher than those provided by Sendust MIG heads. Amorphous Co-based alloys are currently under investigation, since they are the most promising materials for such applications. Co–Fe–Zr–Re amorphous alloys can illustrate the 'tailoring' of materials for applications in thin film perpendicular heads (Wang, Guzman & Kryder, 1990). First, Co–Fe alloys were selected, since they provide the highest saturation magnetisation of all materials (see the Slater–Pauling plot, Fig. 6.3); the alloy was prepared by sputtering onto a cold substrate, to obtain an amorphous film, which resulted in an extremely small coercive field as a result of appropriate annealing; a small amount of Zr was added as glass-former to increase the amorphous phase stability; finally, since Co alloys are characterised by a high magnetostriction, a small amount of Re (rhenium, $Z = 75$) was added to reduce it. An additional advantage of amorphous alloys is their increased corrosion resistance. A schematic head is shown in Fig. 6.35.

Further improvements in the frequency response of thin film heads have been investigated by preparing *multilayered* films with insulating (SiO_2) interlayers (Louis, Jeong & Walser, 1990), or with 'heteroamorphous' structures, where amorphous, magnetic nanoparticles (~ 5 nm) are scattered

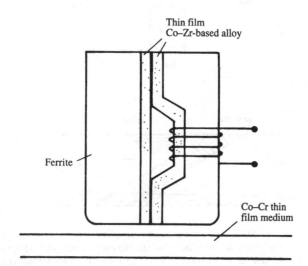

Fig. 6.35. Thin-film head for perpendicular recording (schematic).

in an insulating, amorphous SiO_2 matrix (Matsuyama, Eguchi & Karamon, 1990). This field is currently one of intense activity.

6.4 Materials for magnetooptical recording

Commercial magnetooptical recording systems were introduced in 1990 (Kryder, 1990). This technology is based on thin films with perpendicular anisotropy; information is written by creating a magnetic domain with inverse magnetisation with the help of a high-power laser beam, in an otherwise uniformly magnetised medium, as briefly discussed in Section 5.4.5. Readout is performed by a low-power laser beam, and uses the Kerr effect (or Faraday rotation) to detect the '0' and '1's. Both the writing and reading processes can be repeated virtually endlessly.

The requirements for magnetooptical materials are listed in Table 6.8. Only a few materials can satisfy all these needs. Most magnetooptical films currently used are amorphous rare-earth–transition-metal (R–M) films prepared by sputtering onto glass substrates. In amorphous films, a perpendicular anisotropy can be achieved with the appropriate deposition conditions. Since there is no need for subsequent heat treatments, substrates are pregrooved; these grooves are used as a tracking path for the optical heads. Typical R–M systems are GdTb–Fe and Tb–FeCo (Hansen, 1990).

Storage density is effectively higher in magnetooptical systems than in magnetic recording media; another advantage is that, since the writing and reading are performed by means of light, there is no need to position the head especially close to the film, therefore avoiding the possibility of a 'head crash', as in magnetic recording. A shortcoming, however, is that access times are considerably longer, since two runs are needed to rewrite, or *overwrite* a disc, i.e., to store new information on a previously written disc. The electromagnet used to apply the inverting field usually has a high inductance, and cannot be switched at the data writing rate. In the first run, the field has a constant polarity and data are erased by means of a laser of constant intensity. During the second run, the laser intensity is pulsed and domains of inverted polarity are written (with the field inverted with respect to the first run) whenever the laser power is high enough. Special efforts are currently being made to have a *direct overwrite* system, with access time competitive with those of magnetic recording.

In addition to garnets (see Section 5.4.5), great efforts have been made to prepare new magnetooptical materials involving other R–M systems,

Table 6.8. *Requirements for magnetooptical materials.*

Large uniaxial, perpendicular anisotropy
Large coercive field at room temperature, with a strong temperature dependence
Curie or compensation temperature between 400 and 600 K
Rectangular hysteresis loop
High optical absorption
High Faraday or Kerr rotation angle
Long-term chemical stability

Adapted from Hansen (1990) and Schoenes (1992).

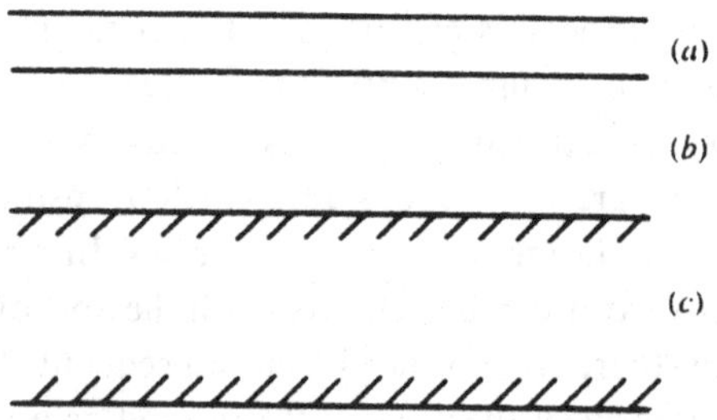

Fig. 6.36. Structure of a magnetooptical thin film: (*a*) dielectric layer; (*b*) high Kerr rotation film; and (*c*) substrate.

such as Pd–Co, Pt–Co and multilayers. A dielectric or reflective layer can enhance the Kerr rotation angle, Fig. 6.36(*a*) (Reim & Weller, 1988; Sato & Kida, 1988). To satisfy the requirements in Table 6.8, a bilayer can be used, made of a high-coercivity, square-hysteresis-loop material, coupled to a low-coercivity, high-Kerr-rotation film, Fig. 6.36(*b*) (Gambino, 1991). The former is used to store the information, based on its high coercivity, and the latter plays the role of a magnetooptical transducer. Pt–Co and Pd–Co multilayers consist of Co layers separated by Pt (or Pd) interlayers. The Co provides a high magnetic moment and the Co–Pt (or Co–Pd) interfaces lead to strong perpendicular anisotropy (Krishnan *et al.*, 1991).

A method for direct overwrite uses two coupled R–M layers. One layer is known as the *reference* layer and has low coercive force and high Curie temperature. The second layer, known as the *memory* layer, has high coercivity and low T_C. First, the magnetisation of the reference layer is oriented downward, by an initialising field with an intensity lower than the coercive field of the memory layer; the latter is therefore unaffected, Fig. 6.37. An upward bias field is applied, with an intensity lower than the value needed to produce an inversion by the field alone. A '0' is written by means of a laser with sufficient energy to reach the T_C of the memory layer, but not enough to overcome the Curie point of the reference layer. On cooling, the memory layer shows a domain with the same orientation as the reference layer at that point. To write a '1', the laser intensity is strong enough to reach the T_C of the reference layer; the magnetisation of both layers is therefore inverted (Kryder, 1990). Impressive improvements can be expected in this area; however, advances in magnetic recording may make it difficult to obtain a larger share of the market for such magnetooptical recording devices.

6.5 *Special magnetic materials and applications*

In this section, a short overview of magnetic materials in an early stage of development and of novel analysis techniques involving magnetic materials is given. It provides a picture of the exciting diversity of magnetic materials.

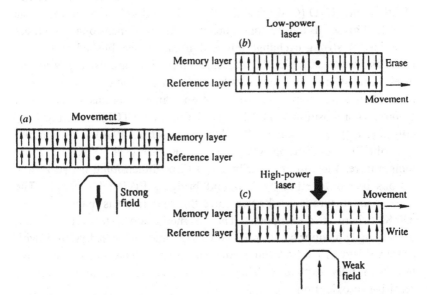

Fig. 6.37. Direct overwrite by beam modulation on a bilayer: (*a*) the reference layer is oriented by a strong field; (*b*) the memory layer is oriented at the low-power laser; and (*c*) both layers are oriented by a weak field at the high-power laser. (Adapted from Marchant, 1990.)

6.5.1 Molecular magnets

The idea of ordered spins within the same molecule seems, in principle, quite exotic. However, exchange is only one particular aspect of atomic bonding between paramagnetic atoms or ions. Instead of needing a three-dimensional network, exchange can take place within a molecule; typically this involves two or more 3d atoms. A three-dimensional solid can then be formed with these molecules. A particular advantage which can be expected is that all the properties and synthesis knowledge developed in organic chemistry can be used to prepare new magnetic materials with unusual properties.

Some of the potential advantages of molecular magnetic materials include the 'tuning' of properties by manipulation of radicals; biocompatibility for drug synthesis in many applications; solubility in organic solvents, facilitating their preparation; use in the form of small aggregates or colloidal dispersions (Gatteschi & Sessoli, 1992).

Currently, the most serious shortcoming for applications of molecular magnets is their extremely low Curie (or Néel) temperatures. The highest reported transition temperature is ~ 30 K, in MnCu[oxamido(N-benzoato-N'-ethanoato]H_2O (Codjovi *et al.*, 1992). Typical values are in the range 1–10 K. This seems to be a consequence of the low dimensionality of these materials. A strong exchange interaction can be established within the molecule; however, molecule–molecule interactions are usually weak. As a result, most molecular magnets behave as a one-dimensional ferro-, or more often, antiferromagnet. An exception is an amorphous material with empirical composition $V(TCNE)_x \cdot yCH_2Cl_2$, where TCNE = tetracyanoethylene, with $x \approx 2$ and $y \approx 1/2$ (Manríquez *et al.*, 1991). This material was obtained by adding $V(C_6H_6)_2$ to an excess of TCNE at room temperature, and can be described as a three-dimensional network of V cations linked together by N-bound bridging [TCNE]$^-$ ligands. The Curie temperature is not known since this material was magnetic up to 350 K, where it decomposes. The magnetisation exhibited a linear, decreasing behaviour between 1.4 and 350 K, and magnetic hysteresis with coercive field of 4.8 kA/m at room temperature. These properties have not been fully explained. The material was unstable and extremely sensitive to oxidation.

Experimentally, the existence of a magnetically ordered phase is often verified by a fitting of the Curie–Weiss relationship (see Section 4.2.3); i.e., by measuring the magnetic susceptibility in the paramagnetic phase.

A good fitting can lead to an estimate of the magnetic moment per molecule, as well as evidence of a ferro-, or an antiferromagnetic arrangement. The reason is that, since the transition temperature can be as low as a few kelvins, it may be extremely difficult to observe clear evidence of a magnetisation curve, or a magnetisation *vs* T curve.

Synthesis techniques can be divided into three classes (Gatteschi & Sessoli, 1992): *organic*, if all the involved paramagnetic radicals are organic; *organic–inorganic*, if metal complexes and organic paramagnetic centres are used; and *inorganic*, if only metal complexes react to produce the molecular magnet. The first approach, using only organic radicals is the most difficult, and only a few molecules have been reported (Fujita *et al.*, 1990). An interesting example is the so-called NTDIOO ferromagnet [2-(4-nitrophenyl)-4,4,5,5-tetramethyl-4,5-dihydro-1H-imidazolyl-1-oxy-3-oxide], which exhibits ferromagnetic behaviour without the presence of transition metals in the organic molecule, Fig. 6.38 (Wan, Wang & Zhao, 1992).

The other two approaches have been more successful. One-dimensional ferrimagnets have been obtained in chains where two different metal ions alternate (Kahn *et al.*, 1988). Some alcohols have shown two-dimensional characteristics; $Mn(SCN)_2(CH_3OH)_2$ can be described as a two-dimensional network of Mn ions bridged by SCN groups with CH_3OH separating the layers. It is weakly ferromagnetic below 10 K, and its magnetic data are

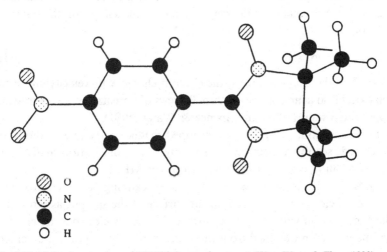

O   O
O   N
●   C
○   H

Fig. 6.38. Molecular structure of NTDIOO organic magnet (Wan, Wang & Zhao, 1992).

consistent with a two-dimensional material (DeFotis *et al.*, 1988). An interesting ionic compound is $[Cr(H_2O)(NH_3)_5][FeCl_6]$, in which discrete $[Cr(H_2O)(NH_3)_5]^{3+}$ and $[FeCl_6]^{3-}$ ions form a three-dimensional lattice held together by ionic forces and H bonds; $T_C = 2.32$ K (Morón *et al.*, 1992). The data could be fitted on the assumption of a non-collinear ferrimagnet.

Once the shortcomings of the low transition temperatures have been solved (presumably by strong, intermolecular interactions in three-dimensional materials), molecular magnets can be expected to contribute to many applications, most likely as magnetooptic materials, where a high flux density is not the basic property. Exciting developments can be foreseen in the near future in this new area.

6.5.2   Magnetic imaging

The observation of magnetic domain structures is important not only for basic studies, but also for the development of magnetic and magnetooptic recording technologies. Some of the observation methods are based on the forces produced by the magnetic fields emerging from the sample, such as the Bitter (briefly described in Section 4.3.1) and magnetic force microscope methods. Other methods make use of the interaction between magnetisation and visible light, such as the Kerr and Faraday techniques (see Section 4.8). Magnetic domains in thin films can be observed by transmission electron microscopy in a defocused mode, or Lorentz microscopy. In this method (also described in Section 4.3.1), the contrast results from the Lorentz force, i.e., the force produced by the magnetic flux of the sample on the travelling electron:

$$\mathbf{F} = q\mathbf{v} \times \mathbf{B} \tag{6.13}$$

where $\mathbf{F}$ is the Lorentz force, q the electron charge, $\mathbf{v}$ its velocity and $\mathbf{B}$ the total local field acting on the electron. Reviews of Lorentz electron microscopy appear in Tsuno (1988) and Spence & Wang (1991).

In this section, two recently developed magnetic imaging methods are described: scanning electron microscopy with polarisation analysis, or SEMPA, and magnetic force microscopy, or MFM.

SEMPA is based on the same technique as scanning electron microscopy: a beam of electrons is focused on the surface of the sample; as a result of their interaction with the surface atoms, secondary electrons are emitted; an image is then obtained from these electrons. The important difference is that, in ferromagnetic samples, the secondary electrons preserve their spin orientation. A spin polarisation analyser can then be used to determine the magnetisation direction and magnitude, since spin polarisation is usually directly related to magnetisation (Scheinfein *et al.*, 1990). To determine the three magnetisation components, two orthogonal spin detectors are used. Magnetisation profiles can be obtained with a resolution of 70 nm. An advantage of this technique is that there is no interaction between the magnetic fields of the sample and the probe, as can occur in methods based on magnetic forces.

The *scanning tunnelling microscope*, STM (Binning *et al.*, 1982), makes use of the tunnelling effect; in short, this is a quantum-mechanical phenomenon (see, for instance, Wolf (1985)) in which electrons from one metal have a non-negligible probability of passing through an insulating barrier to another metal, if this barrier is thin enough; the probability (and therefore the current) is related to the product of electron densities of both metals. In the STM, an extremely fine metal tip is scanned over the surface of the sample, Fig. 6.39; the tunnelling current is kept constant by varying the distance, z, between the tip and the surface. By plotting z as a function of the tip position, an image of the surface topography is generated; from this image, individual atoms can be resolved. The surface of graphite is shown in Fig. 6.40. Another example is the reconstruction of the (111) surface of Si (Binning *et al.*, 1983). A review of this technique appears in García (1991).

A variation of the STM for insulating materials is the *atomic force*

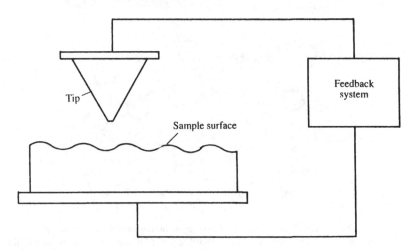

Fig. 6.39. Principle of the STM (schematic).

microscope, AFM. In this technique, the images are formed by measuring the forces on a sharp stylus created by the proximity to the surface of the sample. The displacements of the cantilever are measured by tunnelling, or by means of a capacitive or optical detector; a feedback loop maintains a constant force between the stylus and the surface. The interaction can be measured as a static displacement of the cantilever, or by the deviation from resonance of the cantilever oscillation. Lateral and vertical resolutions of 30 Å and 1 Å , in air, have been obtained (Binning, Quate & Gerber, 1986); a review appears in Hues *et al.* (1993).

The magnetic structure in ferromagnetic materials can be imaged by the magnetic version of AFM, the *magnetic force microscope*, MFM. In this instrument, the cantilever has a ferromagnetic, single-domain tip and

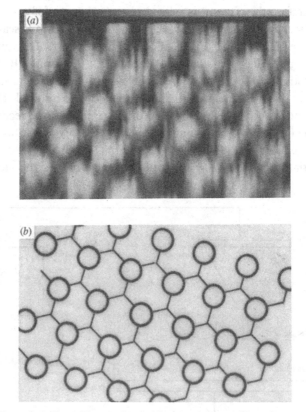

Fig. 6.40. The surface of graphite, as observed by scanning tunnelling microscopy (courtesy of M. Huerta and R. Escudero, University of Mexico); (*b*) reconstruction of the atomic layer.

images are formed by magnetostatic interactions on the surface of the sample. Again, the instrument is operated at constant force by a feedback loop, Fig. 6.41. Usually, the resolution is good, but quantitative studies are difficult. One reason is that parasitic contributions resulting from the mutual deformation of both the tip and the sample configuration lead to complex images. This problem can be avoided by operating at a sufficiently large distance. A second problem is related to the magnetic configuration of the probe, which is, *a priori*, unknown. One solution is to fabricate tips with well-defined symmetry, for instance, a spherical Ni particle deposited on a W tip (Hartmann, 1990). A theoretical approach, relating the geometrical parameters of the probe to the observed contrast has been given by Wadas & Grütter (1989). MFM has been used to study the magnetic structure in Co–Cr thin films with perpendicular anisotropy (Grütter *et al.*, 1989), magnetite (Wiesendanger *et al.*, 1992), and Nd–Fe–B permanent magnets (Grütter *et al.*, 1990), as well as other magnetic materials.

6.5.3 Non-destructive evaluation

A significant part of this book has been devoted to understanding how the microstructure determines magnetic properties. When this knowledge has been developed sufficiently, the opposite relationship can be extremely useful: many particular aspects of the microstructure can be evaluated

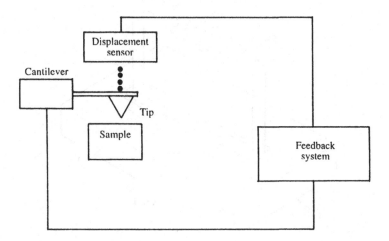

Fig. 6.41. MFM (schematic).

from magnetic properties measurements. These *non-destructive evaluation* methods (NDE for short) for magnetic materials are based on the interaction between defects and the magnetic structure. Magnetic properties are generally very complex, which has resulted in slow development of magnetic NDE methods in the past. However, advances in understanding magnetism coupled with evidence that other methods (such as x-ray inspection, or ultrasonic techniques) are reaching their limits, are leading to an increased interest in magnetic NDE methods. Their importance can be appreciated by considering the massive use of steels (ferromagnetic, iron-rich alloys) in structures, pipelines, containers, etc. Magnetic NDE methods allow not only the *in situ* detection of cracks, but under certain circumstances, can lead to the prediction of failures due to mechanical fatigue and creep.

Cracks, flaws and other defects can be detected by means of *magnetic particle inspection* (MPI). In this method, the piece to be inspected is magnetised by a convenient device, usually a coil, Fig. 6.42; the presence of cracks is detected by applying a magnetic particle suspension (just as in the Bitter method for visualising domain walls, see Section 4.3.1) onto the surface of the piece. Magnetic particles are attracted by any magnetic flux inhomogeneity, such as cracks. In this method, only surface defects are detected. The best detection conditions occur when the largest

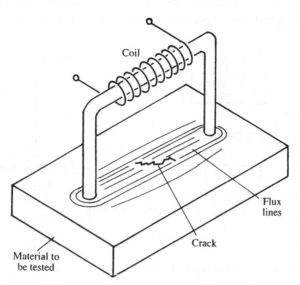

Fig. 6.42. Magnetisation of the area to be tested by MPI (schematic).

dimension of the crack is perpendicular to the applied field. An improvement of this method is to use a fluorescent suspension and to examine the object under ultraviolet light, to enhance any disturbance of the magnetic flux. In complex geometries, the 'replica' method can be useful; in this method, the piece to be inspected is first magnetised; a magnetic tape is then placed on the surface, where it is magnetised by the surface magnetic flux. After removal, the tape is analysed with a magnetometer for any magnetic anomaly. Optimum conditions for this method have been reviewed by Massa (1976).

A method related to MPI is *magnetic flux leakage*. Instead of a magnetic particle suspension, a magnetometer is used to scan the surface of the piece or structure to be tested. Its disadvantage with respect to MPI is that more time is usually needed to inspect a given surface; its advantage is that a more precise measurement of magnetic disturbances can be made, which therefore results in better characterisation of the cracks (size, depth, etc.). The sensitivity of this method also allows an evaluation of the flux anomalies related to the stresses (Jiles, 1991).

Another group of magnetic NDE methods capable of providing an evaluation of grain size, residual stress, fatigue and creep effects, etc., is based on the *Barkhausen effect*. As described briefly in Section 4.3.2, the Barkhausen effect is associated with discontinuous changes in the magnetisation curve, Fig. 6.43. It is related to the sudden unpinning and irreversible displacement of domain walls as a result of increasing field pressure. A Barkhausen plot shows the number of pulses in a given

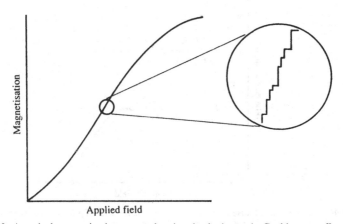

Fig. 6.43. A typical magnetisation curve showing, in the inset, the Barkhausen effect.

field interval, as a function of field strength, Fig. 6.44. The Barkhausen plot depends on the density, distribution and nature of all the defects capable of producing domain wall pinning: point defects, dislocations, grain boundaries, second phases, etc. Since stress affects magnetisation through magnetostriction, the Barkhausen signal also depends on the stress state of the material. A review of these methods appears in Jiles (1991).

The Barkhausen plot is usually obtained in an ac regime, at a given frequency of the exciting signal; magnetic flux variations are extremely small and difficult to detect in dc fields. Since eddy currents act as a shielding against the applied field, the field penetration is limited to a certain depth, which is inversely proportional to frequency. By obtaining the Barkhausen plot at various frequencies, Mayos, Segalini & Putignani (1987) carried out a study of the surface precipitation (carburisation) of a second phase in steels.

Prediction of failure by fatigue in steels can be carried out by Barkhausen effect measurements (Govindaraju *et al.*, 1993). A correlation was observed between the various stages of fatigue (which is damage due to plastic deformation under cyclic load) and the Barkhausen plot. The latter exhibited a maximum between stages one and two, and a rapid increase just before failure. The first stage was characterised by a redistribution of dislocations, which led to an increase in the Barkhausen signal; the discontinuous change in magnetisation increased as the mean distance between pinning sites (dislocations) increased. In the second

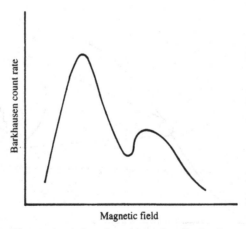

Fig. 6.44. A typical Barkhausen signal plot, where the number of pulses is represented as a function of the applied field (schematic).

stage, the Barkhausen signal decreased; this was attributed to the fact that dislocations accumulated at the surface, as a consequence of their multiplication. The final increase in Barkhausen signal was interpreted on the basis of an increase in domain wall density; as macroscopic cracks were produced, closure domains were nucleated, leading to a significant increase in the total number of domain walls.

Creep damage in steels has been investigated (Devine & Jiles, 1992). Creep can be defined as the slow plastic flow occurring at a high, constant temperature under constant load. It is a serious problem in many industries, where structures and pipelines are subjected to load at a high temperature. Plastic flow during creep results in a build-up of dislocations accompanied by the creation of vacancies. As many vacancies coalesce into voids, cavities and macroscopic cracking occur, leading to failure. The creep process was monitored by coercive field measurements; a decrease in this property was observed to indicate creep damage. This effect was attributed to a decrease in domain wall pinning, as a result of impurity segregation during creep.

6.6   Superconductors

Superconductors are materials with incredible properties: below a certain temperature, known as the critical temperature, T_C, they are both *perfect conductors* (their electrical resistivity is zero) and *perfect diamagnets* (their magnetic susceptibility is -1). Superconductivity was discovered by H. Kamerlingh Onnes in 1911, when the liquefaction of He (4.2 K) opened the possibility of carrying out studies at such low temperatures.

Zero resistivity means that huge current densities can be transported with no energy loss. The simple model for resistivity as a consequence of electron scattering by the crystal lattice, impurites and defects, is clearly unable to account for superconductivity. To explain these properties, Bardeen, Cooper & Schrieffer (1957) assumed that two electrons can form an associated state (a 'bound' state) through an attractive interaction. Electrons normally experience an electrostatic repulsion, since they have the same charge sign; however, if the interaction occurs via the lattice, in the form of a small deformation of the positive ions near the electrons, Fig. 6.45, a small, positive charge concentration attracts a second electron. This *Cooper pair*, as it is known, has two electrons with opposite momenta and spins; as a pair, the total momentum and spin are zero. From the quantum-mechanical point of view, a Cooper pair is represented by a single, coherent wave function with zero wavevector. A zero wavevector involves an infinite wavelength, which, in turn, is interpreted as the absence of scattering, and therefore, has zero resistivity. The distance over which two electrons can be associated, known as the correlation length, ξ, can be evaluated; values are usually in the micrometre range (10^3 nm). Electrons in a Cooper pair are therefore $\sim 10^3$ lattice spacings apart. This (oversimplified!) is the basis of the BCS theory, that successfully explained the properties of *some* superconductors.

The magnetic properties of superconductors are also unusual. When a superconductor is cooled through its critical temperature in the presence of a magnetic field, H, magnetic flux lines are completely expelled from the sample. This is the *Meissner* effect, characteristic of a perfect diamagnetic material ($\chi = 1$). Surface currents are generated in the superconductor to create a magnetic field which exactly opposes the applied field and maintains flux lines outside the material. As H is reduced, surface currents decrease and vanish for $H = 0$. There is, however, a critical value H_c; if the applied field is larger than H_c, the field penetrates the superconductor and it becomes a normal material. The critical field value seems to be related to the maximum current density for the specific superconductor.

There are several types of superconductor, with substantial differences in behaviour. The BCS theory accounts only for properties of materials known as Type I superconductors.

Type I superconductors are usually pure metals with low critical temperatures, Table 6.9. The T_C value depends on the crystal structure, pressure and impurities. In the superconducting stage, Type I super-

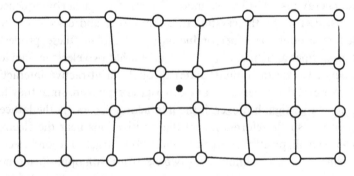

Fig. 6.45. Schematic deformation of the positive-ion lattice by the presence of an electron, in a metallic superconductor.

Table 6.9. *Critical temperature*
for some Type I superconductors.

Superconductor	T_C (K)
Pb	7.2
Ta	4.5
Hg	4.15
In	3.4
Al	1.2
Ga	1.1
Zn	0.8

conductors are extremely sensitive to external magnetic fields; the critical temperature decreases as the applied field, H, increases. There is a certain critical value of field, H_c, above which the superconducting state is destroyed. A 'phase diagram' can be obtained by plotting H vs T, Fig. 6.46, where the superconducting phase exists for low temperatures and low fields. Critical fields are relatively low (in the region of ~ 80 kA/m) for Type I superconductors. The maximum current densities (usually expressed in A/cm^2) in these materials are small.

Type II superconductors are characterised by the fact that a *mixed* state can exist; for certain values of applied field, a partial penetration of field occurs, leading to a mixed or 'vortex' state, which is a combination of normal filaments with finite electrical resistance (where magnetic flux lines penetrate) surrounded by a superconducting matrix, with zero electrical resistance and no flux lines. There are therefore *two* critical field values, H_{c_1} and H_{c_2}: the material is a superconductor for $H < H_{c_1}$, and no field penetration is observed; normal behaviour occurs for $H > H_{c_2}$, with total magnetic flux penetration; and finally, the vortex state exists for $H_{c_1} < H < H_{c_2}$, Fig. 6.47, with partial flux penetration. The 'vortices' are usually associated ('pinned') with microstructural defects such as dislocations, second phases, grain boundaries, etc. When the applied field is increased, vortices are unpinned and grow through the sample, leading to a normal state.

Type II superconductors are intermetallic compounds, Table 6.10, with critical temperatures and upper critical fields considerably higher than Type I materials. This latter characteristic has given rise to superconducting magnets which generate enormous magnetic fields, currently the most important application of superconductors (see applications below).

Until 1985, Nb_3Ge was the superconductor with the highest critical temperature (23 K). In 1986, Bednorz & Muller discovered that a La, Ba and Cu oxide in polycrystalline (or ceramic) form exhibited superconductivity

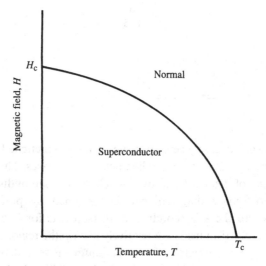

Fig. 6.46. Schematic phase diagram of a Type I superconductor.

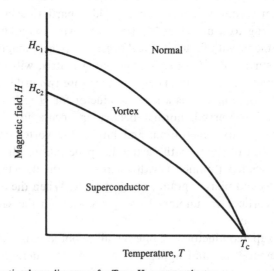

Fig. 6.47. Schematic phase diagram of a Type II superconductor.

Table 6.10. *Critical temperature for some Type II superconductors.*

Superconductor	T_C (K)
Nb_3Ge	23.2
$Nb_3(AlGe)$	20.7
Nb_3Ga	20
Nb_3Al	18.7
Nb_3Sn	18
NbTi	10
V_3Si	16.9
V_3Ga	14.8

Table 6.11. *Critical temperatures of high-temperature superconductors.*

Superconductor	T_C (K)	Reference
Reported as: $Ba_xLa_{5-x}Cu_5O_{5(3-y)}$ $x = 1, 0.75, y > 0$ now known to be: $La_{2-x}Sr_xCuO_4$	30	Bednorz & Müller, April 1986
$YBa_2Cu_3O_x$	92.5	Wu *et al.*, February 1987
$Bi_2Sr_2CaCu_2O_x$	108	Maeda *et al.*, January 1988
$Tl_2Ba_2Ca_2Cu_3O_{10}$	125	Parkin *et al.*, March 1988
Superstructure containing: $HgBa_2Ca_2Cu_3O_{8+x}$ and $HgBa_2CaCu_2O_{6+x}$	133	Schilling *et al.*, May 1993

at temperatures as high as 30 K. The exciting discovery of super-conductivity in such oxides stimulated an extraordinary activity in many laboratories in the world; in just a few years, the highest critical temperature increased from 30 K to 125 K for a Tl, Ba, Ca, Cu oxide (Parkin *et al.*, 1988), and to 133 K, in an ordered superstructure containing $HgBa_2Ca_2Cu_3O_{8+x}$ and $HgBa_2CaCu_2O_{6+x}$ (Schilling *et al.*, 1993), Table 6.11. Critical temperatures higher than liquid nitrogen boiling temperature (77 K) open the possibility for relatively inexpensive operation of superconducting devices since liquid nitrogen is relatively cheap (less than $1/litre).

The crystal structures of *high-temperature superconductors* (HTS) are all modifications of the perovskite structure (prototype: $CaTiO_3$). Most of the HTSs exhibit a layered structure. $YBa_2Cu_3O_\delta$ (also known as '123'), for example, consists of Cu–O planes and Cu–O chains, Fig. 6.48. These structures are anisotropic: electrical conductivity parallel to the basal plane is usually two orders of magnitude higher than in the normal direction. The superconducting mechanism is associated with Cu–O planes, since $Bi_2Sr_2CaCu_2O_\delta$ (also known as 'BiSCCO', or '2212') and $Tl_2Ba_2Ca_2Cu_3O_\delta$, with very high T_C values, have no Cu–O chains (Jiles, 1991).

The O content in HTS, indicated by δ, can be a non-integral number and has a crucial influence on the critical temperature. In '123', T_C drops from 90 K for $\delta \approx 7$, to ~ 60 K for $\delta \approx 6.6$, and then to ~ 20 K for $\delta = 6.3$

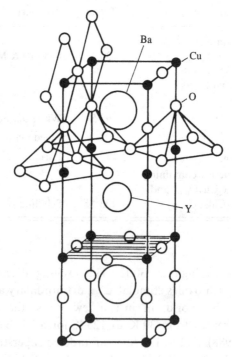

Fig. 6.48. Unit cell of $YBa_2Cu_3O_{7-x}$ ceramic superconductor. (Adapted from Jorgensen *et al.*, 1988.)

(Tarascón *et al.*, 1987). The properties of HTSs cannot be explained by BCS theory; currently, there is not even a well-established basis to account for these properties.

High-T_C superconductors have generated many expectations concerning their applications, ranging from inexpensive energy distribution, to ultra-fast computers and levitated mass transport. However, their ceramic nature makes them extremely brittle and very difficult to produce as a flexible wire, which is the most convenient form to fabricate coils and create high magnetic fields. Another consequence is that these materials usually melt incongruently; their preparation as high-quality single crystals is therefore extremely difficult. As polycrystals, they present many defects (grain boundaries) that limit the critical current density.

Some efforts to produce commercial wires and ribbons are overcoming these difficulties, by the so-called powder-in-tube approach (Amato, 1993). In short, a hollow Ag cylinder is charged with the powdered precursor of a HTS (such as BiSSCO, for example). This composite is pulled through a die of smaller diameter, or flattened by rolling into a ribbon; the Ag sheath deforms easily, exerting a compression on the powder and resulting in an alignment of grains. Once in its final form, the composite is heated to produce the HTS. By this technique, wires 100 m in length with critical current densities in the region 12 000 A/cm^2 have been obtained. This represents a significant step toward commercial applications of HTS.

In 1984, during attemps to reproduce the long-chain C molecules formed in interstellar space by vaporisation of graphite under laser irradiation, a research group in Sussex found an exceptionally stable cluster of 60 C atoms (Kroto *et al.*, 1985). Its structure was a remarkable *spherical* molecule, consisting of a polygon of 60 vertices, 32 faces (20 hexagonal and 12 pentagonal), Fig. 6.49. A C atom on each vertex has two single bonds and one double bond, therefore satisfying all its valencies. These molecules (also known simply as C_{60}) were named *buckminster-fullerenes*, after the American architect R. Buckminster Fuller, who designed a geodesic dome with the same fundamental symmetry. There is, however, a more familiar model for this molecule, with exactly the same symmetry: the football, and therefore, they are also known as *bucky balls*. The reason for including C_{60} molecules in this overview is that, besides being considered by some authors as an entirely new field of science (Culotta & Koshland, 1991), they are superconductors.

Bucky balls combine a series of amazing characteristics: they are extremely stable, but with a wide chemical versatility: C_{60} can react with

Table 6.12. *Critical temperature of some superconducting fullerenes.*

Composition	T_C	Reference
K_3C_{60}	19.3	Stephens *et al.* (1991)
Rb_xC_{60} (x ≈ 3)	28	Rosseinsky *et al.* (1991)
Cs_xC_{60} (x = 1.2 − 3)	30	Kelty, Chen & Lieber (1991)
$Cs_xRb_yC_{60}$ (x = 2, y = 1)	33	Tanigaki *et al.* (1991)

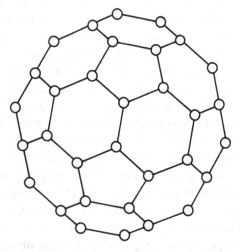

Fig. 6.49. Schematic structure of the buckminsterfullerene C_{60} molecule.

halogens, free radicals, alkali metals, etc. Its physical properties are extremely interesting; its resistance to shock could lead to its use as a lubricant. It has exceptional electronic properties: pure C_{60} is an insulator, but when doped with some K to give K_3C_{60}, it becomes a superconductor. Addition of more K to give K_6C_{60} makes it an insulator again. By playing with a related form, C_{70}, and with various dopants (Rb, Cs, Tl), the critical temperature can be raised from 18 to ~45 K (Culotta and Koshland, 1991), Table 6.12. It was given the 'Molecule of the Year Award' by *Science* in 1991 (Koshland, 1991).

Superconductors

Buckminsterfullerenes can be easily prepared by a simple method, involving the evaporation of graphite electrodes in a He atmosphere (Krätschmer *et al.*, 1990). Because of their versatility, there is some expectation that bucky balls can offer a good model for establishing a mechanism and theory for superconductivity. Bucky balls are not expected to represent a serious competition to ceramic HTS materials, since the critical temperatures of the former are rather low.

The most important application of conventional Type II superconductors (such as Nb_3Ti) is currently the generation of intense, stable magnetic fields in devices such as magnetic resonance imaging (or MRI), particle accelerator fusion research and other devices. MRI is the single largest application of superconductors with some 4000 machines in use in the world (Amato, 1993). This diagnostic technique is based on the resonance properties of nuclei with non-zero spin (odd number of protons), usually H. Protonic spins behave in a similar way to electronic spins (see Section 4.6.2); Eq. (4.68) also applies to nuclear resonance. To obtain a collective response, all the nuclear magnetic moments are first aligned by means of a high, uniform magnetic field (200–450 kA/m) (Kaufman & Crooks, 1985) produced by the superconducting magnet. A set of coils generates linear magnetic field gradients in the three cartesian directions, providing a spatial encoding to nuclear magnetic resonance signals, since resonance conditions depend on the local magnetic field. A third separate set of coils provides an RF signal and detects the absorption of these signals by the sample.

The signal received from the nuclei depends on many parameters, such as their position in the field, the time to recover equilibrium after the disturbance, the time during which they provide a signal, their concentration, their position within the molecule to which they belong, etc. The image formation can be performed with one or several of these parameters, resulting in significant differences in contrast. For example, H from 'fat' can be distinguished from H in water, within the same tissue (Kaufman & Crooks, 1985). Also, some lesions are more clearly visible by using one specific parameter than others. MRI can be used in an extremely wide range of diagnoses, since practically any tissue can be imaged.

Another current application of superconductors is superconducting quantum interference devices or SQUIDs. These devices can measure magnetic flux with an extremely high sensitivity, which is particularly useful for many purposes. The basis of such a high sensitivity is the

297

quantum nature of superconductivity. The formation of Cooper pairs, according to the BCS theory, represents a long-range order (a macroscopic quantum state) with some important consequences. To sustain the supercurrent in a closed path (supercurrent in a loop), the phase of the Cooper pairs must change by $2\pi n$ each turn, where n is an integer. Therefore, the magnetic flux generated by a supercurrent loop cannot take arbitrary values, but is quantised in units of the flux quantum, Φ_0; Φ_0 is the ratio of Planck's constant to the charge of the Cooper pair:

$$\Phi_0 = h/2e = 2.067 \times 10^{15} \text{ Wb} \tag{6.14}$$

A supercurrent can *tunnel* (see Section 6.5.2) from one superconductor to another one, separated by a thin insulating barrier (a *Josephson* junction). In the case of a superconductor junction, however, the phase difference on the two sides of the junction is related to the supercurrent (Clarke, 1986). When two Josephson junctions are connected in parallel to form a superconducting loop, Fig. 6.50, a change in magnetic flux threading the loop results in an oscillation of the critical current with a period equal to Φ_0 (Jaklevic *et al.*, 1964). These oscillations are a consequence of the *interference* between the macroscopic wave functions at the junctions; they allow a change in magnetic flux to be detected as a change in voltage with high accuracy.

The device described above, formed with two Josephson junctions subjected to a constant current, is known as a dc SQUID. Due to progress in thin film technology, SQUIDs with a single Josephson junction, excited

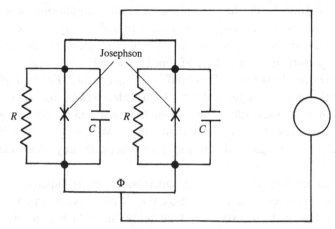

Fig. 6.50. A loop containing two Josephson junctions for a dc-SQUID device (schematic).

by an ac current have been developed. The Josephson junction is coupled to the inductor of a resonant (LC) circuit; the changes in magnetic flux in the inductor lead to changes in the amplitude of the ac voltage (at the resonance frequency) with a period of Φ_0.

SQUIDs have been used in a large number of experiments where the detection of small magnetic fields is crucial. Some interesting cases include the study of magnetic fields from human brains associated with mental disorders (Barth *et al.*, 1984), the detection of gravitational waves (Clarke, 1986), and fluctuations in the earth's magnetic field related to geothermal effects (Clarke, 1983), etc.

References

Allia, P. & Vinai, F. (1987). Losses, aftereffect and disaccommodation in amorphous ferromagnetic alloys. In *Magnetic Properties of Amorphous Metals*. Eds A. Hernando, V. Madurga, M. C. Sánchez-Trujillo and M. Vázquez. North-Holland, Amsterdam, pp. 347–53.

Altounian, Z., Chan, X., Liao, L. X., Ryan, D. H. & Ström-Olsen, J. O. (1993). Structure and magnetic properties of rare earth iron nitrides, carbides and carbonitrides. *Journal of Applied Physics*, **73**, 6017–22.

Altúzar, P. & Valenzuela, R. (1991). Avrami and Kissinger theories for crystallization of metallic amorphous alloys. *Materials Letters*, **11**, 101–4.

Amato, I. (1993). New superconductors: A slow dawn. *Science*, **259**, 306–8.

Bardeen, J., Cooper, L. N. & Schrieffer, J. R. (1957). Microscopic theory of superconductivity. *Physical Review*, **106**, 162–3.

Baró, M. D., Clavaguera, N. & Suriñach, S. (1988). The crystallization process of $Ni_{78}Si_8B_{14}$ amorphous alloys. *Materials Science and Engineering*, **97**, 333–6.

Barth, D. S., Sutherling, W., Engel, J. Jr & Beatty, J. (1984). Neuromagnetic evidence of spatially distributed sources underlying epileptiform spikes in the human brain. *Science*, **223**, 293–6.

Bate, G. (1980). Recording materials. In *Ferromagnetic Materials*, Vol. 2. Ed. E. P. Wohlfarth, North-Holland, Amsterdam, pp. 381–507.

Bednorz, J. G. & Müller, K. A. (1986). Possible high T_C superconductivity in the Ba–La–Cu–O system. *Zeitschrift fur Physik*, **B64**, 189–93.

Binning, G., Rohrer, H., Gerber, Ch. & Weibel, E. (1982). Tunnelling through a controllable vacuum gap. *Applied Physics Letters*, **40**, 178–80.

Binning, G., Rohrer, H., Gerber, Ch. & Weibel, F. (1983). 7 × 7 reconstruction on Si(111) resolved in real space. *Applied Physics Letters*, **50**, 120–3.

Binning, G., Quate, C. F. & Gerber, Ch. (1986). Atomic force microscope. *Physical Review Letters*, **56**, 930–3.

Blanke-Bewersdorff, M. & Köster, U. (1988). Transient nucleation in zirconium-based metallic glasses. *Materials Science and Engineering*, **97**, 313–16.

Bozorth, R. M. (1951). *Ferromagnetism*. Van Nostrand, New York.

Other magnetic materials

Bozorth, R. M. (1953). The permalloy problem. *Review of Modern Physics*, **25**, 42–8.

Burke, J. (1965). *The Kinetics of Phase Transformations in Metals*. Pergamon Press, New York.

Buschow, K. H. J. (1986). New permanent magnet materials. *Materials Science Reports*, **1**, 1–64.

Buschow, K. H. J. (1988). Permanent magnet materials based on 3d-rich ternary compounds. In *Ferromagnetic Materials*, Vol. 4. Eds. E. P. Wohlfarth and K. H. J. Buschow. North-Holland, Amsterdam, pp. 1–130.

Buschow, K. H. J. (1991). New developments in hard magnetic materials. *Reports on Progress in Physics*, **54**, 1123–213.

Cadogan, J. M., Gavigan, J. P., Givord, D. & Li, H. J. (1988). A new approach to the analysis of magnetization measurements in rare earth-transition metal compounds – application to $Nd_2Fe_{14}B$. *Journal of Physics F*, **18**, 779–87.

Cahn, R. W. (1990). Diffusion mechanisms in amorphisation by solid state reaction. *Colloque de Physique*, Colloque C4, **51**, C4-3–10.

Calka, A. & Radlinski, A. P. (1988). The local value of the Avrami exponent: A new approach to devitrification of glassy metallic ribbons. *Materials Science and Engineering*, **97**, 241–6.

Cargill III, G. S. (1975). Structure of metallic alloy glasses. *Solid State Physics*, **30**, 227–320.

Cartoceti, A. (1985). The impact of Nd–Fe–B on the typical applications of permanent magnets. In *Nd–Fe Permanent Magnets: Their Present and Future Applications*. Ed. I. V. Mitchell. Elsevier, London, pp. 209–14.

Chen, C-C. & Lieber, C. (1991). Isotope effect and superconductivity in metal-doped C_{60}. *Science*, **259**, 655–8.

Chen, C. W. (1977). *Magnetism and Metallurgy of Soft Magnetic Materials*. North-Holland, Amsterdam, p. 309.

Chen, C. W. (1977a). *Magnetism and Metallurgy of Soft Magnetic Materials*. North-Holland, Amsterdam, p. 241.

Chikazumi, S. (1964). *Physics of Magnetism*. Wiley, New York.

Chin, G. Y. (1971). Processing control of magnetic properties of magnetostrictive transducer applications. *Journal of Metals*, **23**, 42.

Chin, G. Y. & Wernick, J. H. (1980). Soft metallic materials. In *Ferromagnetic Materials*, Vol. 1. Ed. E. P. Wohlfarth. North-Holland, Amsterdam, pp. 55–158.

Clarke, J. (1983). Geophysical applications of SQUIDs. *IEEE Transactions on Magnetics*, **MAG-19**, 288–94.

Clarke, J. (1986). SQUIDs, brains and gravity waves. *Physics Today*, March 1986, 36–44.

Codjovi, E., Bergerat, P., Nakatani, K., Pei, Y. & Kahn, O. (1992). Molecular-based magnets studied with an ultra-sensitive SQUID magnetometer. *Journal of Magnetism and Magnetic Materials*, **104–107**, 2103–4.

Coey, J. M. D. & Sun, H. (1990). Improved magnetic properties by treatment of iron-based rare earth intermetallic compounds in ammonia. *Journal of Magnetism and Magnetic Materials*, **87**, L251–4.

References

Coey, J. M. D. & Hurley. (1992). New interstitial rare-earth iron intermetallics produced by gas phase reaction. *Journal of Magnetism and Magnetic Materials*, **104–107**, 1098–101.

Cook, J. S. & Rossiter, P. L. (1989). Rare-earth iron boron supermagnets. In *CRC Critical Reviews in Solid State and Materials Sciences*, Vol. 15. CRC Press, Florida, pp. 509–50.

Cornelison, S. G. & Sellmyer, D. J. (1984). Rare earth gallium iron glasses. 1. Magnetic-ordering and hysteresis in alloys based on Gd, Tb and Er. *Physical ReviewB*, **30**, 2845–56.

Croat, J. J., Herbst, J. F., Lee, R. W. & Pinkerton, F. E. (1984). Pr–Fe and Nd–Fe based materials: A new class of high-performance permanent magnets. *Journal of Applied Physics*, **55**, 2078–82.

Cullity, B. D. (1972). *Introduction to Magnetic Materials*. Addison-Wesley, Massachusetts..

Culotta, E. & Koshland, D. E. Jr (1991). Buckyballs: Wide open playing field for chemists. *Science*, **254**, 1706–9.

DeFotis, G. C., McGhee, E. M., Echols, K. R. & Wiese, R. S. (1988). Magnetic and structural properties of $Mn(SCN)_2(ROH)_2$ compounds. *Journal of Applied Physics*, **63**, 3569–71.

Devine, M. K. & Jiles, D. C. (1992). Effects of high temperature creep on magnetic properties of steels. *IEEE Transactions on Magnetics*, **28**, 2465–6.

Duwez, P. (1967). The structure and properties of alloys rapidly quenched from the liquid state. *Transactions of the American Society for Metals*, **60**, 607–16.

Escobar, M. A., Yavari, R., de Lacheisserie, E. T. & González, J. (1991). Saturation magnetostriction of amorphous tapes with $\lambda_s > 0$ and $\lambda_s \sim 0$ after relaxation by conventional and rapid dynamic current annealing. *Materials Science and Engineering*, **A113**, 184–7.

Ferguson, E. T. (1958). Uniaxial magnetic anisotropy induced in Fe–Ni alloys by magnetic anneal. *Journal of Applied Physics*, **29**, 252–3.

Fujimori, H., Yoshimoto, H. & Masumoto, T. (1981). Anomalous eddy current loss and amorphous magnetic materials with low core loss. *Journal of Applied Physics*, **52**, 1893–8.

Fujita, I., Teki, Y., Takui, T., Kinoshita, T., Itoh, K., Miko, F., Sawaki, Y., Iwamura, H., Izuoka, A. & Sugawara, T. (1990). Design, preparation and electron spin resonance detection of a ground state undecet ($S = 5$) hydrocarbon. *Journal of the American Chemical Society*, **112**, 4047–55.

Gambino, R. J. (1991). Exchange coupled films for magnetooptic applications. In *Science and Technology of Nanostructured Magnetic Materials*. Eds. G. C. Hadjipanayis and G. A. Prinz. NATO ASI Series B: Physics, Vol. 259. Plenum Press, London, pp. 91–8.

García, N. (1991). Scanning tunneling microscopy and force microscopy applied to magnetic materials. In *Science and Technology of Nanostructured Materials*. Eds. G. C. Hadjipanayis and G. A. Prinz. NATO ASI Series B: Physics, Vol. 259. Plenum Press, London, pp. 301–30.

Gatteschi, D. & Sessoli, R. (1992). Molecular based magnetic materials. *Journal of Magnetism and Magnetic Materials*, **104–107**, 2092–5.

Other magnetic materials

Gibbs, M. R. J. (1990). Anisotropy and magnetostriction in amorphous alloys. *Journal of Magnetism and Magnetic Materials*, **83**, 329–33.

Gibson, M. A. & Delamore, G. W. (1988). Surface crystallization in melt-spun metallic glasses. *Journal of Materials Science*, **23**, 1164–70.

Gignoux, D. (1992). Magnetic properties of metallic systems. In *Electronic and Magnetic Properties of Metals and Ceramics*, Part I. Ed. H. K. J. Buschow. VCH Publishers, New York, pp. 367–456.

Goertz, M. (1951). Iron–silicon alloys heat treated in a magnetic field. *Journal of Applied Physics*, **22**, 964–5.

Gong, W. & Hadjipanayis, C. G. (1993). Mechanically alloyed 1:12 nitrides and carbides, *Journal of Applied Physics*, **73**, 6245–7.

Gould, J. E. (1959). Progress in permanent magnets. *Proceedings of the Institution of Electrical Engineers*, **106A**, 493–500.

Govindaraju, M. R., Strom, A., Jiles, D. C., Biner, S. B. & Chen, Z-J. (1993). Evaluation of fatigue damage in steel structural components by magnetoelastic Barkhausen signal analysis. *Journal of Applied Physics*, **73**, 6165–7.

Graham, C. D. Jr (1959). Magnetic annealing. In *Magnetic Properties of Metals and Alloys*, American Society for Metals, Cleveland, pp. 288–329.

Grüter, P., Wadas, A., Meyer, E., Hidber, H.-R. & Guntherodt, H.-J. (1989). Magnetic force microscopy of a CoCr thin film. *Journal of Applied Physics*, **66**, 6001–6.

Grütter, P., Jung, Th., Heinzelmann, H., Wadas, A., Meyer, E., Hidber, H.-R. & Guntherodt, H.-J. (1990). 10 nm Resolution by magnetic force microscopy on FeNdB. *Journal of Applied Physics*, **67**, 1437–41.

Guyot, M. & Globus, A. (1976). Determination of the domain wall energy from hysteresis loops in YIG. *Physica status solidi* (b), **59**, 447–54.

Gwan, P. B., Scully, J. P., Bingham, D., Cook, J. S., Day, R. K., Dunlop, J. B. & Heydon, R. G. (1987). Magnetic measurements in rapidly quenched and hot pressed NdFeB alloys. In *Proceedings of the Ninth International Workshop in Rare Earth Magnets*. Eds. C. Hergnet and R. Poerschke. DPG-GmbH, Bad Honnef, pp. 295–9.

Hadjipanayis, G. C., Yadlowski, E. J. & Wollins, S. H. (1982). A study of magnetic hardening in $Sm(Co_{0.69}Fe_{0.22}Cu_{0.07}Zr_{0.02})_{7.22}$. *Journal of Applied Physics*, **53**, 2386–8.

Hansen, P. (1990). Magneto-optical recording materials and technology. *Journal of Magnetism and Magnetic Materials*, **83**, 6–12.

Hartmann, U. (1990). Magnetic force microscopy. *Advanced Materials*, **2**, 550–2.

Hasegawa, R. (1991). Glassy alloy identification marker. *Journal of Applied Physics*, **69**, 5025–6.

Hayashi, K., Hayakawa, M., Ohmori, H., Okabe, A. & Aso, K. (1990). CoPtB(O) alloy films as new perpendicular recording material. *Journal of Applied Physics*, **67**, 5175–7.

Herbst, J. F., Croat, J. J., Pinkerton, F. E. & Yelon, W. B. (1984). Relationship between crystal structure and magnetic properties in $Nd_2Fe_{14}B$. *Physical Review B*, **29**, 4176–8.

302

References

Herzer, G. (1991). Magnetization process in nanocrystalline ferromagnets. *Materials Science and Engineering*, **A133**, 1–5.

Hilzinger, H. R., Mager, A. & Warlimont, H. (1978). Amorphous ferromagnetic materials – magnetic fundamentals, properties and applications. *Journal of Magnetism and Magnetic Materials*, **9**, 191–9.

Hinz, G. & Voigt, H. (1989). Magnetoelastic sensors. In *Magnetic Sensors*. Eds. R. Boll and K. J. Overshott. VCH Publishers Inc., New York, pp. 97–152.

Houze, G. L. (1967). Domain wall motion in grain-oriented silicon steel in cyclic magnetic fields. *Journal of Applied Physics*, **38**, 1089–96.

Hues, S. M., Colton, R. J., Meyer, E. & Guntherodt, H-J. (1993). Scanning probe microscopy of thin films. *Materials Research Society Bulletin*, **XVII**, 41–9.

Iwasaki, S. (1984). Perpendicular magnetic recording – evolution and future. *IEEE Transactions on Magnetics*, **MAG-20**, 657–68.

Jagielinski, T. & Egami, T. (1985). Fictive temperature during the relaxation of field induced anisotropy in amorphous alloys. *IEEE Transactions on Magnetics*, **MAG-21**, 2002–4.

Jaklevic, J. C., Lambe, J., Silver, A. H. & Mercereau, J. E. (1964). Quantum interference effects in Josephson tunneling. *Physical Review Letters*, **12**, 159–62.

Jiles, D. (1991). *Introduction to Magnetism and Magnetic Materials*. Chapman and Hall, London, p. 388.

Jorgensen, J. D., Shaked, H., Hinks, D. G., Dabrowski, B., Veal, B. W., Paulikas, A. R., Nowicki, L. J., Grabtree, G. W., Kwok, W. K., Numez, L. M. & Claus, H. (1988). Oxygen vacancy ordering and superconductivity in $YBa_2Cu_3O_{7-x}$. *Physica*, **C153–155**, 578–81.

Kahn, O., Pei, Y., Verdaguer, M., Renard, J. P. & Sletten, J. (1988). Magnetic ordering of Mn II Cu II bimetallic chains – design of a molecular-based ferromagnet. *Journal of the American Chemical Society*, **110**. 782–9.

Kaufman, L. & Crooks, L. E. (1985). NMR imaging. *Journal of Applied Physics*, **57**, 2989–95.

Kelty, S. P., Chen, C-C. & Lieber, C. (1991). Superconductivity at 30 K in caesium-doped C_{60}. *Nature*, **352**, 223–4.

Kobayashi, T., Kubota, M., Satoh, H., Kumura, T., Yamauchi, K. & Takahashi, S. (1985). A tilted sendust sputtered ferrite video head. *IEEE Transactions on Magnetics*, **MAG-11**, 1536–8.

Kogure, T., Katayama, S. & Ishii, N. (1990). High-coercivity magnetic hard disks using glass substrates. *Journal of Applied Physics*, **67**, 4701–3.

Koshland, D. E. Jr (1991). Molecule of the year. *Science*, **254**, 1705.

Köster, U. (1988). Surface crystallization of metallic glasses. *Materials Science & Engineering*, **97**, 233–9.

Krätschmer, W., Lamb, L. D., Fostiropoulos, K. & Huffman, D. R. (1990). Solid C_{60}: a new form of carbon. *Nature*, **347**, 354–8.

Krishnan, R., Lassri, H. & Rougier, P. (1987). Magnetic properties of amorphous FeErBSi ribbons. *Journal of Applied Physics*, **62**, 3463–4.

Krishnan, R., Porte, M., Tessier, M. & Flevaris, N. K. (1991). Structural and magnetic studies in Co–Pt multilayers. In *Science and Technology of Nanostructured Magnetic Materials*. Eds. G. C. Hadjipanayis and G. A.

Prinz. NATO ASI Series B: Physics Vol. 259. Plenum Press, London, pp. 191–4.

Kroto, H. W., Heath, J. R., O'Brien, S. C., Curl, R. F. & Smalley, R. E. (1985). C_{60}: Buckminsterfullerene. *Nature*, **318**, 162–3.

Kryder, M. H. (1990). Advances in magneto-optic recording technology. *Journal of Magnetism and Magnetic Materials*, **83**, 1–5.

Lee, R. W., Brewer, E. G. & Schaffel, N. A. (1985). Processing of neodymium-iron-boron melt-spun ribbons to fully densed magnets. *IEEE Transactions on Magnetics*, **MAG-21**, 1958–63.

Louis, E., Jeong, I. S. & Walser, R. M. (1990). Bias and frequency response of the permeability of $CoZrNb/SiO_2$ multilayers. *Journal of Applied Physics*, **67**, 5117–19.

Luborsky, F. (1980). Amorphous ferromagnets. In *Ferromagnetic Materials*, Vol. 1. Ed. E. P. Wohlfarth, North-Holland, Amsterdam, pp. 451–530.

Maeda, H., Tanaka, Y., Fukutomi, M. & Asano, T. (1988). A new high-T_C oxide superconductor without a rare-earth element. *Japanese Journal of Physics 2: Letters*, **27**, L209–10.

Maletta, H. & Zinn, W. (1989). Spin glasses. In *Handbook on the Physics and Chemistry of Rare Earths*, Vol. 12. Eds. H. A. Gschneider and L. Eyring. North-Holland, Amsterdam, pp. 213–356.

Mallinson, J. C. (1987). *The Foundations of Magnetic Recording*. Academic Press, London.

Manríquez, J. M., Yee, G. T., McLean, R. S., Epstein, A. J. & Miller, J. S. (1991). A room-temperature molecular organic-based magnet. *Science*, **252**, 1415–17.

Marchant, A. B. (1990). *Optical Recording – A Technical Overview*. Addison-Wesley, Massachusetts, p. 95.

Marseglia, E. A. (1980). Kinetic theory of crystallization of amorphous materials. *Journal of Noncrystalline Solids*, **41**, 31–6.

Massa, G. M. (1976). Finding the optimum conditions for weld testing by magnetic particles. *Nondestructive Testing International*, **9**, 16–26.

Masumoto, T., Watanabe, K., Mitera, M. & Ohnuma, S. (1977). High magnetic permeability amorphous alloys of the FeNiSiB system. In *Amorphous Magnetism II*. Eds. R. A. Levy and R. Hasegawa, Plenum Press, New York, pp. 369–77.

Matsuyama, H., Eguchi, H. & Karamon, H. (1990). The high-resistive soft magnetic amorphous films consisting of cobalt, iron, boron, silicon and oxygen, utilized for video head devices. *Journal of Applied Physics*, **67**, 5123–5.

Mayos, M., Segalini, S. & Putignani, M. (1987). Electromagnetic nondestructive evaluation of surface decarburization on steels, *Mémoires et Etudes Scientifiques de la Revue de Métallurgie*, France, **85**, 85–96.

McCurrie, R. A. (1982). The structure and properties of alnico permanent magnet alloys. In *Ferromagnetic Materials*, Vol. 3. Ed. E. P. Wohlfarth. North-Holland, Amsterdam, pp. 127–88.

McCurrie, R. A. & Willmore, L. E. (1979). Barkhausen discontinuities of

References

nucleation and pinning of domain walls in etched microparticles of $SmCo_5$. *Journal of Applied Physics*, **50**, 3560-4.

McGuiness, P. J., Harris, I. R., Rozendaal, E., Ormerod, J. & Ward, M. (1986). The production of a NdFeB permanent magnet by hydrogen decrepitation/ attribution milling route. *Journal of Materials Science*, **21**, 4107-10.

Mishra, R. K., Chu, T. Y. & Rabenberg, L. K. (1990). The development of the microstructure of die-upset NdFeB magnets. *Journal of Magnetism and Magnetic Materials*, **84**, 88-94.

Morón, M. C., Palacio, F., Pons, J. & Casabó, J. (1992). Magnetic properties of the low temperature ferrimagnet $[Cr(H_2O)(NH_3)_5][FeCl_6]$. *Journal of Magnetism and Magnetic Materials*, **114**, 243-5.

Morrish, A. H. (1965). *The Physical Principles of Magnetism*. Wiley, New York, p. 279.

Mydosh, J. A. & Nieuwenhuys, G. J. (1980). Dilute transitional metal alloys: Spin glasses. In *Ferromagnetic Materials*, Vol. 1. Ed. E. P. Wohlfarth. North-Holland, Amsterdam, pp. 71-182.

Nozawa, T., Matsuo, Y., Kobayashi, H., Iayama, K. & Takahashi, N. (1988). Magnetic properties and domain structures in domain refined grain-oriented silicon steel. *Journal of Applied Physics*, **63**, 2966-70.

O'Handley, R. C., Hasegawa, R., Ray, R. & Chou, C. P. (1976). Ferromagnetic properties of some new metallic glasses. *Applied Physics Letters*, **29**, 330-2.

Ormerod, J. (1985). The physical metallurgy and processing of sintered rare earth permanent magnets. *Journal of Less Common Metals*, **1**, 49-69.

Parkin, S. S. P., Lee, V. Y., Engler, E. M., Nazzal, A. I., Huang, T. C., Gorman, G., Savoy, R. & Beyers, R. (1988). Bulk superconductivity at 125 K in $Tl_2Ca_2Ba_2Cu_3O_x$. *Physical Review Letters*, **60**, 2539-42.

Ramanan, V. R. V. & Fish, G. E. (1982). Crystallization kinetics in Fe-B-Si metallic glasses. *Journal of Applied Physics*, **53**, 2273-5.

Ranganathan, S. & von Heimendhal, M. (1981). The 3 activation energies with isothermal transformations - applications to metallic glasses. *Journal of Materials Science*, **16**, 2401-4.

Ray, A. E., Soffa, W. A., Blachere, J. R. & Zhang, B. (1987). Microstructure development in $Sm(Co, Fe, Cu, Zr)_{8.35}$ alloys. *IEEE Transactions on Magnetics*, **MAG-23**, 2711-13.

Reim, W. & Weller, D. (1988). Kerr rotation enhancement in metallic bilayer thin films for magneto-optical recording. *Applied Physics Letters*, **53**, 2453-4.

Rosseinsky, M. J., Ramírez, A. P., Glarum, S. H., Murphy, D. W., Haddon, R. C., Hebard, A. F., Palstra, T. T. M., Kortan, A. R., Zahurak, S. M. & Makhija, A. V. (1991). Superconductivity at 28 K in Rb_xC_{60}. *Physical Review Letters*, **60**, 2030-2.

Ruigrok, J. J. M., Sillen, C. W. M. P. & van Rijn, L. R. M. (1990). High performance metal-in-gap heads with very small track widths. *Journal of Magnetism and Magnetic Materials*, **83**, 41-4.

Sagawa, M., Fujimura, S., Togawa, N., Yamamoto, H. & Matsuura, Y. (1984). New materials for permanent magnets on a base of Nd and Fe. *Journal of Applied Physics*, **55**, 2083-7.

Other magnetic materials

Saito, J., Sato, M., Matsumoto, H. & Akasaka, H. (1987). Direct overwrite by light power modulation on magnetooptical multilayered media. *Japanese Journal of Applied Physics*, **26**, Supplement 26-4: Proceedings of the International Symposium on Optical Memory '87, 155–9.

Sato, K. & Kida, H. (1988). Calculations of magneto-optical spectra in compositionally-modulated multilayered films. *Journal de Physique*, Colloque C8, C8-1779–80.

Scheinfein, M. R., Unguris, J., Pierce, D. T. & Celotta, R. J. (1990). High spatial resolution quantitative macromagnetics. *Journal of Applied Physics*, **67**, 5932–7.

Schilling, A., Cantoni, M., Guo, J. D. & Ott, H. R. (1993). Superconductivity above 130 K in the HgBaCaCuO system. *Nature*, **363**, 56–7.

Schoenes, J. (1992). Magneto-optical properties of metals, alloys and compounds. In *Electrical and Magnetic Properties of Metals and Ceramics*, Part I. Ed. K. H. J. Buschow. VCH Publishers Inc., New York, pp. 147–257.

Sellmyer, D. J., Wang, D. & Christner, J. A. (1990). Magnetic and structural properties of CoCrTA films and multilayers with Cr. *Journal of Applied Physics*, **67**, 4710–12.

Shibaya, H. & Fukuda, I. (1977). Preparation by sputtering of thick sendust films suited for recording head core. *IEEE Transaction on Magnetics*, **MAG-13**, 1029–35.

Slater, J. C. (1936). The ferromagnetism of nickel. *Physical Review*, **49**, 537–45.

Slonczewski, J. C. (1963). Magnetic annealing. In *Magnetism*, Vol. 1. Eds. G. T. Rado and H. Suhl, Academic Press, New York, pp. 295–42.

Spence, J. C. H. & Wang, Z. L. (1991). Electron microscope methods for imaging internal magnetic fields at high spatial resolution. In *Science and Technology of Nanostructured Magnetic Materials*. Eds. G. C. Hadjipanayis and G. A. Prinz, NATO ASI Series B: Physics Vol. 259. Plenum Press, London, pp. 279–300.

Stanley, J. K. (1963). *Electrical and Magnetic Properties of Metals*. American Society for Metals, Metal Park, Ohio.

Stephens, P. W., Milhaly, L., Lee, P. L., Whetten, R. L., Huang, S.-M., Kaner, R., Deiderich, F. & Holczer, K. (1991). Structure of single-phase superconducting K_3C_{60}. *Nature*, **351**, 632–3.

Strnat, K. J. (1988). Rare earth–cobalt permanent magnets. In *Ferromagnetic Materials*, Vol. 4. Eds. E. P. Wohlfarth and K. H. J. Buschow. North-Holland, Amsterdam, pp. 131–209.

Ström-Olsen, J. O., Ryan, D. H., Altounian, Z. & Brüning, R. (1991). Structural relaxation and the glass transition in a metallic glass. *Materials Science and Engineering*, **A133**, 403–9.

Taguchi, S., Sakakura, A., Matsumoto, F., Takashima, K. & Kuroki, K. (1976). The development of grain-oriented silicon steel with high permeability. *Journal of Magnetism and Magnetic Materials*, **2**, 121–31.

Tanigaki, K., Ebbesen, T. W., Saito, S., Mizuki, J., Tsai, J. S., Kubo, Y. & Kuroshima, S. (1991). Superconductivity at 33 K in $Cs_xRb_yC_{60}$. *Nature*, **352**, 222–3.

References

Tarascón, J. M., McKinnon, W. R., Greene, L. H., Hull, G. W. & Vogel, E. M. (1987). Oxygen and rare earth doping of the 90 K superconducting perovskite $YBa_2Cu_3O_{7-x}$. *Physical Review B*, **36**, 226–34.

Tsumita, N., Shiroishi, Y., Suzuki, H., Ohno, T. & Uesaka, Y. (1987). Perpendicular and longitudinal magnetic properties and crystallographic orientation of Co–Cr film. *Journal of Applied Physics*, **61**, 3143–5.

Tsuno, K. (1988). Magnetic domain observation by means of Lorentz electron microscopy with scanning technique. *Reviews of Solid State Science*, **2**, 623–58.

Valenzuela, R., Huanosta, A., Medina, C., Krishnan, R. & Tarhouni, M. (1982). *In situ* crystallization studies of Co-evaporated amorphous FeZr thin films. *Journal of Applied Physics*, **53**, 7786–8.

Valenzuela, R. & Irvine, J. T. S. (1992). Effects of thermal annealing on the magnetization dynamics of vitrovac amorphous ribbons. *Journal of Applied Physics*, **72**, 1486–9.

Valenzuela, R. & Irvine, J. T. S. (1993). Domain wall dynamics and short range order in ferromagnetic amorphous ribbons. *Journal of Noncrystalline Solids*, **156–158**, 315–18.

Wadas, A. & Grütter, P. (1989). Theoretical approach to magnetic force microscopy. *Physical Review B*, **39**, 12013–17.

Wan, M., Wang, H. & Zhao, J. (1992). Organic ferromagnet of NTDIOO crystal. *Journal of Magnetism and Magnetic Materials*, **104–107**, 2096–8.

Wang, S., Guzman, J. I. & Kryder, M. H. (1990). High moment soft amorphous CoFeZrRe thin film materials. *Journal of Applied Physics*, **67**, 5114–16.

Wang, Y. Z., Hu, B. P., Rao, X. L., Liu, G. C., Yin, L., Lai, W. Y., Gong, W. & Hadjipanayis, G. C. (1993). Structure and magnetic properties of $NdFe_{12-x}Mo_xN_{1-\delta}$ compounds. *Journal of Applied Physics*, **73**, 6251–3.

Warlimont, H. & Boll, R. (1982). Applications of amorphous soft magnetic materials. *Journal of Applied Physics*, **26**, 97–105.

Wiesendanger, R., Shvets, I. V., Bürgler, D., Tarrach, G., Guntherodt, H.-J., Coey, J. M. D. & Gräeser, S. (1992). Topographic and magnetic-sensitive scanning tunneling microscope study of magnetite. *Science*, **255**, 583–6.

Wolf, E. L. (1985). *Principles of Electron Tunneling Spectroscopy*. Oxford University Press, Oxford.

Wu, M. K., Ashburn, J. R., Torng, C. J., Hor, P. H., Meng, R. L., Gao, L., Huang, Z. J., Wang, W. Q. & Chu, C. W. (1987). Superconductivity at 93 K in a new mixed-phase Y–Ba–Cu–O compound system at ambient pressure. *Physical Review Letters*, **58**, 908–10.

Zhao, Z. G., Wang, J. Y., Yang, F. M., Sun, K. X., Chuang, Y. C. & de Boer, F. R. (1993). Magnetic properties of $R_2Co_{14}(BC)$ compounds (R = Y, Sm). *Journal of Applied Physics*, **73**, 5875–7.

Zijlstra, H. (1980). Permanent magnets: theory. In *Ferromagnetic Materials*, Vol. 3. Ed. E. P. Wohlfarth. North-Holland, Amsterdam, pp. 37–106.

Index

Index

Index